Taste and Smell Disorders

RHINOLOGY AND SINUSOLOGY
Diagnosis • Medical Management • Surgical Approaches

Series Editor

Howard L. Levine, M.D., F.A.C.S.
Director, The Mt. Sinai Nasal-Sinus Center
Chief, Section of Nasal-Sinus Surgery
The Mt. Sinai Medical Center
Cleveland, Ohio

Taste and Smell Disorders

Allen M. Seiden, M.D., F.A.C.S.
Department of Otorhinolaryngology—Head and Neck Surgery
University of Cincinnati
Cincinnati, Ohio

1997
THIEME
New York • Stuttgart

Thieme
381 Park Avenue South
New York, New York 10016

TASTE AND SMELL DISORDERS
Allen M. Seiden

Library of Congress Cataloging-in-Publication Data
Taste and smell disorders/[edited by] Allen M. Seiden.
    p. cm. — (Rhinology and sinusology)
    ISBN 0-86577-533-8. — ISBN 3-13-107261-X
    1. Smell disorders. 2. Taste disorders. I. Seiden, Allen M.
II. Series.
    [DNLM: 1. Anosmia. 2. Taste Disorders. WV 301T215 1997]
    RF341.T34 1997
    616.8′7—dc20
    DNLM/DLC
    for Library of Congress   96-23875
                                                   CIP

Printed in the United States of America.

5   4   3   2   1

TMP ISBN 0-86577-533-8
GTV ISBN 3-13-107261-X

Contents

Contributors

Richard M. Costanzo, Ph.D.
Department of Physiology and Neurology
Medical College of Virginia
Richmond, Virginia

Terence M. Davidson, M.D.
Department of Otolaryngology — Head
    and Neck Surgery
University of California School of Medicine
San Diego, California

Laurence J. DiNardo, M.D., F.A.C.S.
Department of Otolaryngology — Head
    and Neck Surgery
Medical College of Virginia
Richmond, Virginia

Richard L. Doty, Ph.D.
Smell and Taste Center
University of Pennsylvania
Philadelphia, Pennsylvania

Heather J. Duncan, Ph.D.
Department of Internal Medicine
University of Cincinnati College of
    Medicine
Cincinnati, Ohio

Pamela M. Eller, M.S.
Department of Otolaryngology — Head
    and Neck Surgery
University of Colorado
Denver, Colorado

Joel B. Epstein, D.M.D., MS.D.
Department of Dentistry
Etobicoke General Hospital
Rexdale, Ontario, Canada

Marion E. Frank, Ph.D.
Chemosensory Clinical Research Center
University of Connecticut Health Center
Farmington, Connecticut

Janneane F. Gent, Ph.D.
John B. Pierce Laboratory
New Haven, Connecticut

Miriam Grushka, M.Sc., D.D.S., Ph.D.
Department of Dentistry
Etobicoke General Hospital
Rexdale, Ontario, Canada

Lloyd Hastings, Ph.D.
Department of Environmental Health
University of Cincinnati College of
    Medicine
Cincinnati, Ohio

Bruce W. Jafek, M.D.
Department of Otolaryngology — Head
    and Neck Surgery
University of Colorado
Denver, Colorado

Edward W. Johnson, Ph.D.
Department of Otolaryngology — Head
    and Neck Surgery
University of Colorado
Denver, Colorado

Marian L. Miller, Ph.D.
Department of Environmental Health
University of Cincinnati College of
    Medicine
Cincinnait, Ohio

April E. Mott, M.D.
Chemosensory Clinical Research Center
University of Connecticut Health Center
Farmington, Connecticut

Claire Murphy, Ph.D.
Department of Psychology
San Diego State University
San Diego, California

Bruce Murrow, M.D., Ph.D.
Department of Otolaryngology — Head
    and Neck Surgery
University of Colorado
Denver, Colorado

William T. Nickell, Ph.D.
Department of Otolaryngology — Head
    and Neck Surgery
University of Cincinnati College of
    Medicine
Cincinnati, Ohio

Jill Razani, M.A.
Department of Psychology
San Diego State University
San Diego, California

Allen M. Seiden, M.D., F.A.C.S.
Department of Otolaryngology — Head
    and Neck Surgery
University of Cincinnati College of
    Medicine
Cincinnati, Ohio

David V. Smith, Ph.D.
Department of Anatomy
University of Maryland School of Medicine
Baltimore, Maryland

James B. Snow, Jr., M.D.
National Institute on Deafness and Other
    Communication Disorders
Bethesda, Maryland

Foreword

It is estimated that about two million Americans have a taste and smell disorder and that about 200,000 of these people visit physicians each year. Almost certainly, there are large numbers of individuals who have smell and taste disorders who do not seek help and therefore are unreported.[1] All of these individuals have a significant disability.

Since the recent growing interest in nasal and sinus disorders, otolaryngologists have started looking at these patients for treatable causes and to be certain that there is no serious etiology. In addition, there is a growing body of basic science and clinical research in taste and smell disorders. In an attempt to satisfy the clinician's needs for a source of information about recent basic research and clinical approaches, this volume, *Taste and Smell Disorders*, has been written. Allen M. Seiden, M.D., F.A.C.S., was chosen as its editor because of his long interest in this area.

Taste and Smell Disorders provides the practitioner with information about basic anatomy, physiology, and biochemistry of taste and smell. It organizes a diagnostic approach which is practical and usable. It considers the differential diagnosis and management for each etiologic entity.

Taste and Smell Disorders is another volume in the ongoing series *Rhinology and Sinusology*. Each of the textbooks in the series is meant to serve as a book to be read and studied or used as a practical reference for all health professionals with interest in this area.

Howard L. Levine, M.D., F.A.C.S.

REFERENCE

1. Smell and Taste Disorders. Washington, DC: American Academy of Otolaryngology—Head and Neck Surgery, Inc., 1986.

Preface

The special senses of taste and smell enable the recognition of chemical signals from our environment and are essential for the survival of most vertebrate species. By contrast, these senses have become almost vestigial in man, who has become visually dominant and no longer requires them for survival. Interestingly, the steps involved in chemoreception bear some resemblance to those involved in photoreception. However, while color vision enables the discrimination of several hundred hues, and is accomplished by the comparative processing of information from only three classes of photoreceptors, a countless variety of olfactory receptors are capable of discriminating over 10 000 structurally distinct odorants.

This remarkable amount of information colors our world in ways that are difficult to quantify. Having had the opportunity to evaluate many patients who have sustained a chemosensory loss, I have become painfully aware of the devastating impact upon their quality of life. Unfortunately, the clinical evaluation of chemosensory disorders has lagged far behind that of the other sensory systems, eg, vision and hearing. Historically, the focus of rhinology also has largely been on nasal physiology and its air conditioning mechanism, as well as disease states that obviate this function, with little attention paid to associated olfactory impairment.

Unbeknownst to most otolaryngologists, there has been a plethora of basic research within the chemical senses over these last two decades. This very active community, united under the auspices of the Association for Chemoreception Sciences, has greatly expanded our understanding of taste and smell physiology. This expanded knowledge base in turn has fostered the development of a number of clinical taste and smell research centers within the United States, combining both basic and clinical research activities. Clinically relevant information is therefore beginning to emerge at an increasing pace.

I was very excited when Howard Levine, M.D. asked me to compile a book on taste and smell disorders as part of the Rhinology and Sinusology series. This series covers a vast amount of material and, as you can see, does not neglect the chemical senses. A number of larger, more comprehensive texts covering taste and smell have recently appeared in print. However, this book is more readable to the practicing otolaryngologist and provides a more clinically focused background with which to better evaluate patients presenting with taste and smell complaints.

The first chapter emphasizes the importance of the chemical senses, and is written by James B. Snow, Jr., M.D. Dr. Snow currently directs the National Institute on Deafness and Other Communication Disorders (NIDCD) at the NIH, and pioneered the complementary relationship between basic scientist and otolaryngologist in an attempt to further explore chemosensory disorders. Chapter 2 then provides an overview of the initial approach to patients presenting with a taste or smell loss.

The remainder of the book is divided between olfaction and gustation. Chapter 3 presents a comprehensive review of the anatomy and physiology of olfaction, while chapter 4 covers techniques of olfactory testing. The importance of utilizing accurate and reproducible testing methods when examining patients with an olfactory loss cannot be overemphasized. Chapter 5 through 8 discuss the most commonly identified causes of olfactory loss, which will account for over 50% of patients presenting with olfactory dysfunction. Chapter 9 then reviews the current state of the art regarding olfactory mucosal biopsy and histopathology, a technique that holds great promise in helping us to better understand the pathophysiology of olfactory loss.

Chapter 10 moves on to discuss the anatomy and physiology of taste, while chapter 11 reviews practical techniques for measuring taste function. While taste loss is not as common in clinical practice as olfactory loss, burning mouth syndrome is a related, very difficult problem that is comprehensively reviewed in chapter 12.

The final chapter discusses aging and the chemical senses. Each of the sensory systems begins to deteriorate with advancing age, and as our aging population increases, this subject will become increasingly important.

All of the contributors to this volume have recognized expertise in the field, and together have presented a detailed but clinically applicable review of the chemical senses. It is hoped that this will not only facilitate evaluation and treatment of these patients, but will ultimately stimulate a broader interest among otolaryngologists to pursue chemosensory disorders in greater depth.

Allen M. Seiden, M.D., F.A.C.S.

To my parents, George and Pearl, who taught me the value of honesty, the way to sincerity, and the importance of education. Their inspiration is a guiding light.

To my wife, Peachy, and three children, Brittany, Chad, and Zachary, who are my center of gravity.

To my secretary, Debbie Herbig, who has stayed with me through it all.

—A.M.S.

Importance of
the Chemical Senses

JAMES B. SNOW, Jr., M.D.

The chemical experiences we describe as smell and taste provide a unique armature for understanding the molecular mechanisms of evolution, inheritance, and reception, perception, and interpretation of sensory information. These powerful senses undergird the human experience of environment.

For many years, the chemical senses had been miscast as mere enhancements that offer an improved quality of life for an individual by contributing to the flavor and palatability of food and beverage and the enjoyment of the fragrances of fruits, flowers, and nuts and all that is manufactured to assimilate their odors. Currently, there is expanding awareness of the role played by smell and taste in providing important disease markers and early warning signals for danger and disease. Abnormalities in smell and taste functions frequently accompany and even signal the existence of several diseases or high-risk conditions including diabetes; hypertension; degenerative diseases of the nervous system such as Parkinson's disease, Alzheimer's disease, and Korsakoff's psychosis; obesity; and malnutrition. Smell and taste alert individuals to life-threatening circumstances such as fires, poisonous fumes, leaking gas, spoiled foods or beverages, or poisonous plants. Smell and taste losses can also lead to dietary indiscretion and serious depression.

Progress in basic research is yielding clinical benefits for those who suffer from diseases and disorders of the chemical senses. Importantly, progress in chemosensory mechanisms can be applied broadly to many other areas of biomedical inquiry. For example, it has recently been demonstrated that repeated exposure to an odorant to which an animal is insensitive results over time in sensitivity to that specific odorant but not to other odorants.[1] Inducing olfactory sensitivity by repeated exposure to an odorant suggests stimulus-induced plasticity in a sensory receptor neuron and a novel form of stimulus-controlled gene expression. Not only does the application of olfactory induction have therapeutic potential in chemosensory disorders, but it may also open a new area of opportunity in basic molecular biology.

Smell and taste cells are the only sensory cells that are regularly replaced throughout the life span. Examination of the phenomena by which these and other damaged sensory cells may be replaced has application to hearing, balance, and vision. Progress in transplantation in the central nervous system occurred when segments of the olfactory neuroepithelium were placed in the brain where the tissue thrived and the stem cells divided, giving rise to new receptor neurons that developed, matured, and connected with target tissue in the brain. Since olfactory nerves have the ability to pass barriers that impede other nerve cells, olfactory transplants have the potential to serve as pathways for the reconnection of injured neurons with their targets. Transplants of olfactory neuroepithelium provide a new approach for the biological repair of nervous system damage resulting from injury, neurodegenerative diseases, and aging.

The phylogeny and ontogeny of the chemical senses and the inheritance of olfactory and gustatory skills, deficits, and predispositions to early loss or chemical destruction of receptor cells are of critical interest to chemosensory investigators. The building blocks of genetic differences are being revealed as well as the emerging and waning abilities of these senses over the life span of an individual.

Taste cells are reconstituted throughout the lifetime, whereas taste buds develop based upon a critical neonatal period of neuron induction. Important keys to understanding the evolving and sustained ability of adult mammals to taste are based on research of age-related increased sensitivity of adult mammals to sodium coupled with knowledge of the peripheral gustatory system's sensitivity to amiloride. These gustatory clues permit examination of the specialization of function of the amiloride-sensitive channel.

Remarkable progress has been made in establishing the nature of changes that occur in the chemical senses with age. The differential toll taken on the sense of smell with age is of continuing interest, as is the greater ability of women of all ages to identify odors more accurately than men. Of special concern are the effects of smell and taste on the nutritional status of both the very young and the very old. Some strategies with premature infants as well as full-term infants have shown enhanced ability to thrive based on chemosensory stimulation. Additionally, research on loss or distortion of the sense of smell can affect the nutritional status of older individuals. With the right interventions, this problem can be ameliorated or reversed.

Of great concern is the destruction visited upon the chemical senses by pollutants that may not only devastate the senses themselves but also use the olfactory system as a portal to reach the brain, bypassing the blood-brain barrier. The nature of the pollutants and their effect on these delicate systems indicate the importance of the underlying molecular basis for chemoreception and the tolerable levels to specific substances.

Humans with normal ability to smell can perceive as many as tens of thousands of odorants. Recent progress on inducing smell sensitivity gives hope that genetic defects related to smell and taste may be correctable. Some two million Americans suffer with disorders of smell and taste. The hereditary nature of chemosensory disorders is just beginning to be revealed. The extremely important discovery of an enormous family of receptor genes has set the stage for rapid progress in olfaction.[2] There may be as many as 1000 olfactory receptor genes. This likens the olfactory system to the immune system in terms of the number of genes dedicated to it. Scientists are attempting to reveal the mechanisms and the organizational principles that serve highly developed discrimination of a huge variety of odorants.

The electric life of the chemosenses provides further insight into processes and mechanisms. Upon receiving an odorant, for example, receptor cells translate that stimulus into an electric response and another electric response is delivered to the brain. It has been through the recent development of ultraprecise electrophysiological techniques that these processes are being better understood. In the area of taste reception, techniques employing synthetic organic chemistry and neurophysiological and biochemical strategies are being used to measure receptor mechanisms.

The clinical application of this electrophysiology in the case of olfaction results in measurements of event-related olfactory potentials. Electrophysiological measurements at the sensory epithelium and more centrally have long been confounded by trigeminal chemical and mechanical sensitivity. Although it is conceptually useful to differentiate between conductive or transport olfactory impairment and sensorineural olfactory impairment because of the goal of making an anatomical diagnosis first so that the number of etiological diagnostic possibilities are reduced, it appears that this goal will have to await the kind of evolution in the sophistication of the interpretation of olfactory-evoked potentials that occurred in auditory-evoked potentials in the 1960s, 1970s, and 1980s and continues today. We do not have the ability to differentiate transport olfactory impairment from sensorineural olfactory impairment, nor can we differentiate sensory from neural olfactory impairments. As facility with olfactory-evoked potentials increases, we can anticipate progress in their use in differential anatomical diagnosis and can even hope for the etiological help that has eluded us in the auditory system.

An area of interest and promise is related to the ways in which the chemical senses may direct or influence behaviors. Pheromones are the chemical signals that influence behaviors. Pheromones are detected by the vomeronasal organ and influence sexual, reproductive, and maternal behaviors in lower animals. Although the vomeronasal organ is present in the human fetus, it was thought to disappear in humans by the time of birth. Recent studies of biopsied human tissue reveal a specialized epithelial organ in the appropriate location whose cells express proteins found in the neurons of the olfactory neuroepithelium. In lower animals, the vomeronasal organ sends messages to the hypothalamus, influencing drives and emotions. While a vomeronasal organ may indeed be present in adult humans, only further study will determine if the function of this structure has any effect on human behavior.

Employing molecular biology, molecular genetics, electrophysiology, and biochemistry and using highly refined tools and measures, scientists working around the world are elucidating the structure and function of the olfactory, gustatory, and trigeminal systems. This will yield improved care for those who suffer disorders of these sensory systems and will contribute to a heightened understanding of the nature and complexity of human interaction with the environment.

REFERENCES

1. Wysocki CJ, Dorries KM, Beauchamp GK: Ability to perceive androstenone can be acquired by ostensibly anosmic people. Proc Natl Acad Sci USA 86:7976–7978, 1989.
2. Buck L, Axel R: A novel multigene family may encode odorant receptors, a molecular basis for odor recognition. Cell 65:175–187, 1991.

2

The Initial Assessment of Patients With Taste and Smell Disorders

ALLEN M. SEIDEN, M.D., F.A.C.S.

It has been estimated that as many as two million Americans suffer from chemosensory dysfunction,[1] yet the relative importance of these disorders continues to be underestimated. Most otolaryngologists will see the occasional patient presenting with a primary complaint of olfactory or gustatory loss, and then may be unsure as to the best procedure for establishing a definitive diagnosis. As a result they may be unable to make any recommendation concerning further therapy and prognosis.

On the other hand, when confronted with a patient with a chemosensory complaint, many clinicians will feel compelled to order a full neuroradiological evaluation. While this does rule out the possibility of intracranial pathology, in most cases this is unnecessary and provides little insight as to the underlying etiology. This underscores the importance of a thorough history and physical examination, which must include some form of chemosensory testing.

In this chapter, I will attempt to provide an overview of chemosensory disorders, introducing a number of points that should be considered when confronted with these clinical problems. The ensuing chapters will cover these issues in much greater detail.

DEFINITIONS

Numerous terms have been used to describe the various manifestations of taste and smell dysfunction, but it is important that terminology be standardized. In this volume, anosmia and ageusia refer to a complete loss of smell and taste, respectively, whereas hyposmia and hypogeusia refer to a partial loss (Table 2–1). These conditions are readily detected by current chemosensory testing methods.

4

Table 2–1   Terms Describing Taste and Smell Dysfunction

Anosmia	Complete loss of smell
Hyposmia	Diminished sense of smell
Hyperosmia	Enhanced odorant sensitivity
Dysosmia	Distortion in odorant perception:
	Parosmia: in response to
	environmental stimulus
	Phantosmia: no stimulus present
Ageusia	Complete loss of taste
Hypogeusia	Diminished taste sensitivity
Hypergeusia	Enhanced taste sensitivity
Dysgeusia	Distortion in taste perception

Specific anosmia refers to the inability to detect only specific odorants, and a patient rarely presents with specific anosmia for clinical evaluation. Hyperosmia describes an increased sensitivity to odors, a not uncommon patient complaint but one that is difficult to objectively assess.

Dysosmia refers to a distorted perception of smell, whether precipitated by an environmental stimulus (parosmia) or occurring spontaneously (phantosmia). This may be intermittent or constant and is almost always unpleasant, frequently described as sewerlike or rancid. It cannot be measured by current techniques, and patients are understandably quite disturbed by this problem. It usually occurs in association with olfactory loss, particularly when viral induced.[2] It has been suggested that distortions of actual odors imply the existence of an intact but impaired olfactory system, while phantom smells may be perceived in a disconnected system.[3] However, previous speculation that dysosmia may represent a sign of regeneration and eventual recovery has not been borne out by further investigation.[4] Nevertheless, one study did find it to be more common in patients with posttraumatic hyposmia rather than anosmia.[5] In the case of sinus pathology, it may be associated with an underlying purulent infection, thereby representing odors actually arising from the sinus cavities. In addition, phantosmias or, in essence, olfactory hallucinations have been described in association with seizure activity, psychiatric illness, and Alzheimer's disease.[6] For a more in-depth discussion of dysosmia, the reader is referred to the excellent chapter by Leopold.[7]

Dysgeusia is a distorted taste perception, is more often persistent, and likewise may or may not be precipitated by an external stimulus. Eating may alternatively exacerbate or mask the symptom. However, in the former case it may be so severe as to overpower any other taste or flavor perception and drastically interfere with proper nutritional intake.[8] Again, the dysgeusia may be described as rancid, but is also frequently characterized as bitter, salty, or metallic. It is not usually associated with a whole mouth taste deficit, although there is evidence to suggest a relationship to spatial ageusia and release of inhibition phenomena.[9] Therefore, it may result from neural injury due to prior dental or oral surgery, or injury to the chorda tympani nerve following otological surgery. Dysgeusia has also been reported to occur secondary to dental disease, the use of certain medications, gastroesophageal reflux, and chronic postnasal drip.[4] However, in many instances it remains idio-pathic, and as such it can be very difficult to treat (see chapter 12).

ETIOLOGY OF TASTE AND SMELL LOSS

More than 200 conditions and 40 medications have been reported to be causative in taste and smell disorders, although much of the information in this regard is anec-dotal.[4,10] While this suggests that a myriad of systemic and neoplastic disorders might ultimately need to be considered when evaluating chemosensory dysfunction, there are a number of more common etiologies that initially should be explored.

Of 339 patients presenting to the University of Cincinnati Taste and Smell Center, 285 (84%) were found to have olfactory loss or complaints of dysosmia[11] (Table 2–2). The most commonly identified causes of olfactory loss in this group of patients were head injury (63 patients, 19%), a prior upper respiratory infection (58 patients, 17%), and nasal or sinus disease (53 patients, 16%), accounting for over 50% of patients seen. A similar clinical experience has been reported by the Con-necticut Chemosensory Clinical Research Center[12] and the University of Pennsylva-nia Smell and Taste Center.[13] Therefore, it is important initially to consider these three possibilities.

A diagnosis of traumatic olfactory loss is based on a history of trauma immedi-ately preceding the loss. Approximately 5% of patients suffering head injury will have an associated olfactory loss, usually involving a frontal or occipital blow, but not necessarily a loss of consciousness[14] (See chapter 7). The presumed mechanism is a shearing of the olfactory filaments as they pass through the cribriform plate, due to coup-contrecoup forces. Preliminary histological study of the olfactory mucosa following traumatic injury has demonstrated marked degenerative changes, suggest-

Table 2–2   Etiology of Smell and Taste Dysfunction
(N = 339)

ETIOLOGY	N	%
Head injury	63	19
Post-URI	58	17
Nasal/sinus disease	53	16
Idiopathic-nasal	58	17
Toxic exposure-nasal	17	5
Multiple	18	5
Congenital	8	2
Age	4	1
Miscellaneous-nasal	6	2
Idiopathic-oral	31	9
Miscellaneous-oral	21	6
Toxic exposure-oral	2	1

URI = Upper respiratory infection. From Seiden.[11]

ing severance of the olfactory nerve (see chapter 9). As might be expected, these patients are typically anosmic, as noted in 71% of patients in a recent study.[11] They tend to be younger, between 20 and 50 years, and male, since this group is more apt to suffer head trauma.[2]

A diagnosis of viral-induced olfactory loss is made when the loss immediately follows an upper respiratory infection, in the absence of any other apparent etiology. Patients clearly describe a persisting olfactory deficit after resolution of cold or flu symptoms, often describing such infections as having been particularly severe (see chapter 6). The precise mechanism of loss in such cases has not been clearly delineated; however, preliminary biopsy studies suggest there is a direct insult to the olfactory neuroepithelium.[15,16] These patients may be anosmic or hyposmic, and as many as two thirds will complain of dysosmia.[11] They tend to be older, often over 60, and for reasons unknown demonstrate a female preponderance of 2:1.[2] Although a viral-induced loss will not fluctuate, many of these patients do experience a very gradual improvement over a period of several years.[17]

Nasal and sinus disease that causes an olfactory loss generally does so by obstructing the nasal vault and olfactory cleft. This may be due to frank polyps, but may also occur from secondary edema due to localized pathology, such as within the ostiomeatal complex, without evidence of intranasal polyps.[18] As such, these patients may not complain of nasal obstruction, and in fact may have little in the way of other sinus symptoms.[11] This can make the diagnosis more difficult, particularly when trying to distinguish from a viral etiology (see chapter 5). The distinction is most important, since in this case the loss is conductive, and the underlying olfactory neuroepithelium remains intact. Therefore, it is the only etiology that is amenable to specific treatment.

Unfortunately, in a fairly substantial number of patients the precise etiology remains undetermined. In the University of Cincinnati study, 17% of patients with olfactory loss were classified as idiopathic.[11] In fact, a number of these patients may well fall into one of the other known diagnostic categories. However, usually due to failings in either the history or in follow-up, such a diagnosis was not clearly accurate. Again, other centers have reported similar experiences.[12,13]

A less common but still significant cause of olfactory loss is exposure to a variety of toxic industrial and environmental agents (Table 2–2). Such episodes are more often related to sudden inadvertent exposure to excessive amounts rather than to low-grade exposure over many years (see chapter 8). Most of these cases occur in the workplace and have included exposure to phosphorous fire, chlorine gas, metal dusts, solvents, acid fumes, oil vapors, and even some household cleaners.[11,19] The specific effect of environmental pollutants on olfactory sensitivity has not been adequately studied.

Table 2–2 lists 18 patients (5%) as having multiple causes simply because although an apparent cause was identified, another etiology could not be ruled out. For example, a patient is suffering a viral infection when suddenly exposed to a toxic substance at work.

Interestingly, eight patients (2%) were noted to have a congenital loss of smell. All reported never having experienced an odorant sensation, and all were found to be anosmic. Although a number of well-described congenital syndromes have been associated with olfactory loss, such as Turner's syndrome and Kallmann's syndrome,[20] none of these patients had such associated characteristics. A preliminary biopsy study has shown the presence of incompletely developed or immature olfactory neuroepithelium in these patients.[21]

Olfactory loss may occur with advancing age, but usually does not become significant until after the seventh decade[22] (see chapter 13). Table 2–2 lists four patients (1%) in whom age was felt to correlate with the degree of olfactory loss. Such a diagnosis must be based upon accurate chemosensory testing that is compared with age-matched controls, in the absence of any other contributing factors. It has been shown that with advancing age, the olfactory neuroepithelium is gradually replaced by respiratory epithelium in a patchy, random distribution.[23] Whether this occurs due to repeated environmental insult or is simply a part of the aging process is unclear, but presumably relates to the associated loss.

Table 2–2 lists six patients (2%) as miscellaneous-nasal and includes those presenting with olfactory loss from a variety of less common causes, such as prior craniofacial surgery or stroke. While such factors will frequently produce a loss of smell, it is unusual that olfactory loss will be the primary complaint.

Although complaints of taste loss are common, as will be discussed below, a measurable loss of taste is distinctly uncommon. This is due to the wider distribution of gustatory receptors that rely on several cranial nerves (see chapter 10), as compared to the olfactory system. Even if a focal deficit should occur with damage to a specific nerve, such a loss is frequently not detected by the patient.[24]

Table 2–2 lists 54 patients of 339 that presented to the University of Cincinnati Taste and Smell Center with a measurable taste loss or complaints of dysgeusia. In 31 patients (9%) the etiology could not be identified. Two patients (1%) presented with taste loss secondary to a toxic exposure, and 21 patients presented with dysgeusia, a focal taste loss, or hypogeusia due to a variety of causes. These included preexisting medications, prior dental surgery, prior otological surgery, and gastroesophageal reflux. Only one patient was found to have a complete ageusia, the etiology of which remained unknown.[11]

Deems et al[13] have reviewed 750 patients presenting to the University of Pennsylvania Smell and Taste Center. Only 23 patients had a measurable taste loss, most often (7 patients, 30%) a result of head trauma. Other identifiable causes included a viral upper respiratory infection (3 patients, 13%), toxic chemical exposure (3 patients, 13%), medications (3 patients, 13%), radiation therapy (1 patient, 4%), and dental procedures (2 patients, 9%). On the other hand, 241 patients had complaints of dysgeusia, most often associated with upper respiratory infection, head trauma, and sinonasal disease, or were otherwise labeled as idiopathic. Of interest is that symptoms of depression were reported to be more prevalent in those patients experiencing dysgeusia and dysosmia.

PATIENT EVALUATION

History

The first step in obtaining a history in these patients is to take the chemosensory complaint seriously and be sympathetic to the patient's problem. Evidence abounds as to the deleterious effect such disorders can have on the quality of life,[7,13] and many of these patients have been to a variety of physicians to try to get help for their problem. Without formal testing, and therefore lack of objective verification of their loss, the validity of their complaint is often questioned not only by health professionals, but also by family and friends. This lack of support adds to the patient's sense of isolation and despair.

Questioning should initially attempt to define the patient's chemosensory complaint, whether it includes a smell loss, taste loss, and/or a distorted perception. It is important to note that patients will often present complaining of both a smell and taste loss, or of primarily a taste loss, when in fact taste function remains intact. Of 750 patients presenting to the University of Pennsylvania Smell and Taste Center, 78.1% complained of a reduced ability to smell and 66.4% of a reduced ability to taste.[13] However, chemosensory testing revealed that while 70.9% indeed had olfactory deficits, less than 3% had a measurable taste deficit. From the Connecticut Chemosensory Clinical Research Center, Goodspeed et al[25] reported only olfactory loss in patients complaining of taste loss.

This discrepancy reflects the common inability of many patients to distinguish between the taste and flavor of foods. The perception of flavor involves olfactory, tactile, and thermal sensations as well as taste, with olfactory input arguably being the most crucial.[26] The loss of such input makes foods in general taste flat and unappealing. Therefore it is much more revealing to ask patients specifically about the four taste qualities, ie, salty, sour, sweet, and bitter, in order to determine whether they truly have a taste deficit.

Inquiry should then be directed toward implicating one of the more common etiologies of chemosensory dysfunction listed above. Therefore, it is particularly important to focus on any antecedent event that may have been associated with symptom onset, such as trauma, a viral upper respiratory syndrome, or any possible exposures. It is important to determine whether onset was acute, as generally occurs after a viral or traumatic insult, or gradual, as might occur with nasal and sinus disease. A slowly progressive loss might, in rare cases, suggest a systemic or neoplastic process. In the case of olfactory loss, it is important to note whether symptoms remain constant or fluctuate, perhaps in response to a variety of physical or environmental activities. If fluctuation does occur, this would suggest an obstructive or conductive loss and clearly implicate nasal or sinus pathology. On the other hand, it must be noted that less than half of patients presenting with a conductive olfactory loss will report a history of fluctuation[11] (see chapter 5).

Several studies indicate that patients suffering an olfactory loss secondary to trauma will tend to be anosmic, and this may be helpful when making a diagnosis.[2,27] However, while earlier reports found that a viral infection and nasal pathology would tend to produce hyposmia and anosmia, respectively,[2] a more recent study found no such difference.[11] In other words, such patients were equally likely to be either anosmic or hyposmic.

On the other hand, dysosmia is reported twice as often in patients suffering a postviral olfactory loss.[11] A complaint of dysosmia should be defined as to whether it is a parosmia or phantosmia, whether it can be masked, and whether it can be detected by others. The latter would be unusual and would suggest an underlying sinus infection in most cases. Rarely it may be associated with an underlying metabolic disorder, such as trimethylaminuria.[28] Along these same lines it may be helpful to note whether the odor is unilateral or bilateral, and whether it is detected on inspiration or expiration.

Any associated systemic symptoms or medical conditions that might be causative need to be evaluated. Naturally a history of chronic sinusitis or allergic rhinitis is important, and the presence of any associated nasal symptoms needs to be determined, as does a history of prior nasal or sinus surgery. It is also important to consider the possibility of both acute and chronic exposures at home and at the workplace. Inquiry should be made regarding other neurological symptoms, as well as any pertinent family history of neurological or medical disease. Olfactory loss has been reported in association with renal failure, liver disease, diabetes, and hypothyroidism, and these conditions need to be considered with a thorough medical history.[29]

It is necessary to delineate the medications a patient is taking or has taken. For example, chemotherapeutic agents can produce a loss of both smell and taste.[30] By the same token, radiation therapy to the head and neck has been shown to cause at least a transient loss of both senses.[31,32]

If a patient is noted to have a taste loss, inquiry needs to be made regarding other oral or dental symptoms, such as dry mouth, burning, or toothache. A history of recent dental work, periodontal disease, or foul odor is important. Associated otological symptoms or surgery might also be significant.

It is important to stress that in the case of chemosensory loss, the history can in many cases be misleading. Patients may have great difficulty defining their symptoms, and certainly may be unable to volunteer pertinent information without being specifically questioned. For that reason, many taste and smell clinics rely on an extensive intake questionnaire that is reviewed in detail with each patient.[26] This also emphasizes the importance of accurate chemosensory testing and a thorough physical examination.

Physical Examination

A complete head and neck examination is essential, while the remainder of the examination is directed largely by the history. When the complaint is olfactory loss, it is very important to distinguish a conductive from a sensorineural etiology. As alluded to above, this cannot always be accomplished by history alone. Even in the absence of other nasal or sinus symptoms, and certainly if the history suggests a postviral loss, a thorough nasal examination is crucial.

Anterior rhinoscopy should evaluate for the presence of rhinitis, whether it be allergic, atrophic, infectious, or vasomotor. Septal deviation alone is unlikely to cause an olfactory loss, and the possibility of further nasal or sinus disease needs to be considered (see chapter 5). This includes underlying infection, polyps, neoplasia, and other signs of inflammation. However, anterior rhinoscopy could in fact be normal despite the presence of pathology obstructing the olfactory cleft, and therefore is not adequate to rule out a conductive olfactory loss.

With the advent of nasal endoscopy, more accurate rhinologic diagnosis is now possible. Although the olfactory cleft is frequently too small to allow direct visualization of the olfactory neuroepithelium, it is possible to detect pathological changes that will obstruct transport of odorant molecules to the nasal vault. The condition of the nasal mucosa and secretions can now be readily assessed. Polyps located high in the nose that otherwise could not be seen, or evidence of ostiomeatal pathology, would suggest a conductive loss. Therefore, nasal endoscopy is an essential part of the examination.

If an olfactory loss cannot be related to nasal or viral pathology, trauma, or some sort of toxic exposure, then a search must be made for another less common etiology. A full neurological examination is important. Again, although it is unlikely that such patients will present with smell loss as their primary complaint, a number of neurological disorders appear to have olfactory loss as an early symptom. These include Alzheimer's disease and Parkinson's disease.[33] Focal neurological findings might suggest a central neoplastic process or stroke.

When patients present with hypogeusia or dysgeusia, a thorough oral and pharyngeal examination is imperative. This includes an assessment of oral hygiene and the condition of the teeth, looking for restorations, prostheses, gingivitis, and halitosis, all of which may be implicated in taste dysfunction.[34] Visual inspection and bimanual palpation are important. Disorders to consider include candidiasis, xerostomia, chronic sialadenitis, and cryptic tonsillitis. A further delineation of dysgeusia as being of central or peripheral origin can be made by the use of a mild topical anesthetic; if application alleviates the foul taste, this would indicate a peripheral source.[35]

Chemosensory Testing

A thorough review of olfactory and gustatory testing is provided in subsequent chapters (chapters 4 and 11), so they will not be covered here in detail. It is simply important to stress that an appropriate method be utilized to accurately document and verify a patient's chemosensory complaint. In the not too distant past, this was clinically impractical. However, as will become apparent from reading these chapters, there are now a variety of methods readily available to the clinician that are reproducible and reliable.

Olfactory testing usually involves identification of specific odors, an assessment of threshold, or both; taste testing usually relies upon the patient's assessment of quality (salty, sour, etc) and intensity, as well as location within the oral cavity.

Due to the frequent confusion of smell, taste, and flavor, as noted above, it is prudent to assess both olfactory and gustatory function in most patients, regardless of their chemosensory complaint. It should be clear that it is not enough to simply ask the patient, nor is it adequate to have him or her identify several random odors or taste solutions. To be meaningful, the evaluation must follow sound psychophysical testing procedures. Therefore, one of the well-established techniques currently in use at the various clinical taste and smell centers is recommended.[27,36]

Radiological Evaluation

Radiological study in many cases can provide valuable clues as to the etiology of chemosensory dysfunction. However, the most common causes can usually be diagnosed by history and physical examination, and therefore x-rays are not ordered routinely. For example, if a patient's history clearly indicates that an upper respiratory infection preceded an olfactory loss and there are no associated neuro-logical signs or symptoms, a scan to rule out intracranial pathology is usually not necessary.

On the other hand, should there be no apparent cause, then the possibility of an undiagnosed intracranial lesion becomes more relevant. An enlarging olfactory groove meningioma, pituitary tumors with suprasellar extension, frontal lobe gliomas, large aneurysms of the anterior cerebral or anterior communicating arter-

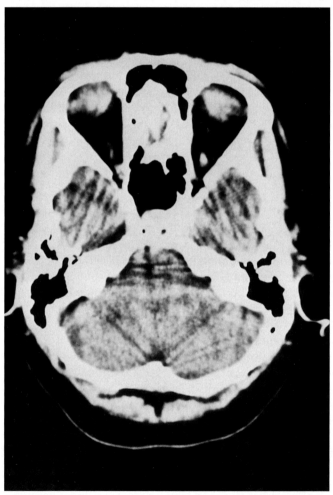

A

Figure 2–1. **A**: Axial head computed tomography (CT) scan obtained in a patient complaining of dysosmia for 8 months. This lower cut captures the upper portion of the ethmoid and sphenoid sinuses. It was interpreted as normal. **B**: Nasal endoscopic exam of same patient after thorough nasal deconges-tion reveals a purulent discharge from the left sphenoethmoidal recess draining into the nasopharynx (arrow). a-adenoid pad, b-left eustachian tube. **C**: An additional CT scan of the paranasal sinuses was obtained in the same patient. This coronal cut through the sphenoid sinus demonstrates complete opacification of the left sphenoid sinus (arrow), consistent with the endoscopic findings.

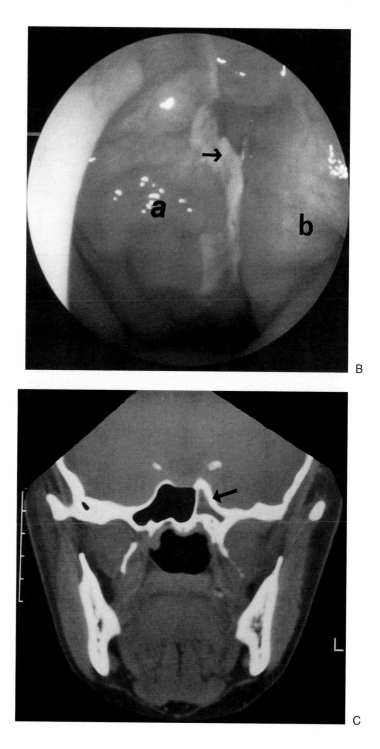

Figure 2–1. *Continued*

ies, and other lesions of the anterior cranial fossa may all produce a gradually progressive olfactory loss. However, as noted previously, such lesions will not generally present with olfactory loss as their only symptom. Magnetic resonance imaging is the procedure of choice when searching for intracranial causes of chemosensory dysfunction.[37]

When a patient presents with olfactory loss that is thought to be conductive, a computed tomographic (CT) scan of the paranasal sinuses is most helpful, preferably in the coronal plane. This provides the clearest detail of both bony and soft tissue anatomy and demonstrates the presence of any underlying inflammatory disease.[38] Proper visualization of the olfactory cleft can be achieved, although even minor degrees of ostiomeatal pathology may be consistent with an obstructive olfactory loss (see chapter 5).

Plain sinus radiographs are generally inadequate for diagnosing a conductive olfactory loss, since they fail to properly delineate the ethmoid sinuses and upper nasal cavity. High-resolution CT scanning is the most definitive way to rule out a conductive etiology. However, in most cases an appropriate decision to order further radiological study can be made based upon a thorough history, chemosensory evaluation, and physical examination that includes nasal endoscopy.

Due to the difficulty frequently encountered trying to determine the etiology of a taste or smell loss, and due to the concern regarding the possibility of unsuspected neoplasia, patients are often sent for radiological study before being referred to an otolaryngologist. In many cases, such head scans will include cuts that incorporate a portion of the paranasal sinuses. However, if the clinician suspects the possibility of underlying sinus disease, these cuts can be unreliable. They may not have been taken low enough to properly evaluate the sinuses, and the window parameters utilized may not be adequate for demonstrating mucosal inflammatory disease. In addition, views in the coronal plane may be more demonstrative of subtle inflammatory changes.

For example, Figure 2–1A depicts the CT scan of a patient complaining of dysosmia for 8 months. This head scan was obtained prior to referral and interpreted as normal, having incorporated the upper portion of the ethmoid and sphenoid sinuses. However, a subsequent nasal endoscopic examination that utilized thorough nasal decongestion revealed a purulent discharge from the left spheno-ethmoidal recess (Fig. 2–1B). An additional CT scan of the paranasal sinuses was ordered and clearly demonstrated opacification of the left sphenoid sinus (Fig. 2–1C). Therefore, should the history or physical examination be suspicious for sinus disease, additional scans may need to be considered.

Additional Diagnostic Tests

The routine use of laboratory tests is rarely helpful, but may be considered should the history dictate. For example, both olfactory and taste loss have been reported in association with hypothyroidism and diabetes mellitus,[29] and further questioning that might elicit associated symptoms could be helpful. In a series of 750 patients presenting to a taste and smell center, Deems et al[13] listed 2 patients having olfactory loss secondary to lupus and 2 secondary to sarcoidosis. Such underlying problems will usually have been previously established by past medical examination but may need to be further explored if no alternative etiologies for chemosensory dysfunction are apparent.

On the other hand, while it has been shown that certain vitamin and mineral deficiencies, such as those of vitamin B_{12}, zinc, or copper, can alter taste or smell function, these cases are extremely rare.[39] Evaluation for such imbalances need not be performed unless clearly indicated by other signs or symptoms, or if by history patients are so predisposed (see below).

In 1982, Lovell, Jafek, and others[40] described an instrument and a technique for taking a biopsy of the olfactory mucosa. Although prior biopsy reports had appeared in the literature,[15] no other group had taken this approach in such a systematic fashion (see chapter 9). The biopsy material is difficult to prepare, but can potentially yield useful diagnostic information.

Moran et al[41] examined two patients suffering a posttraumatic smell loss and found a reduced number of ciliated receptors with a disorganized-appearing epithelium when viewing by transmission electron microscopy. They postulated that these findings were consistent with severing of the olfactory nerve. Jafek et al[16] examined 17 patients with postviral olfactory dysfunction and found degenerative changes within the neuroepithelium. Histopathological changes have also been described in the case of congenital olfactory loss.[21]

While further histological study promises to enhance our understanding of the pathophysiology of olfactory loss, consistent results can be difficult to achieve. Since it is invasive, and due to the difficulty in obtaining adequate specimens, the delicate nature of the tissue, and the expertise required for its interpretation, at present olfactory mucosal biopsy would appear to remain largely a research tool.

THE ZINC CONTROVERSY

It is quite common for patients suffering a disorder of taste or smell to be placed on a therapeutic trial of zinc. This stems from a widely held notion that zinc deficiency is a common occurrence in idiopathic hypogeusia and hyposmia, and that zinc supplementation may therefore be effective. There are a number of factors responsible for this misconception.

First, there is evidence that zinc is important for proper functioning of the special senses, including vision, taste, and smell.[39] However, although it is an important cofactor for a number of metalloenzymes and is required for a variety of metabolic activities, its precise role in the chemical senses is not clear.[42] Nevertheless, it has been noted that in several disease states in which zinc depletion is common, taste dysfunction may occur.

For example, decreased taste acuity has been reported in association with chronic liver diseases including alcoholic cirrhosis and viral hepatitis, renal failure, and Crohn's disease, all of which can predispose to a deficiency of zinc.[39] In addition, taste abnormalities have been found in patients taking a variety of systemic chelating agents, such as penicillamine, that increase the urinary excretion of zinc and copper.[42] While such evidence is circumstantial, this led a number of investigators to supplement zinc in such patients and measure its effect upon taste.

Mahajan et al[43] conducted a double-blind study in 22 hemodialysis patients randomized to receive either zinc acetate supplements or placebo. After 6 to 12 weeks, taste testing demonstrated normalization of both detection and recognition thresholds for the group receiving zinc, but no effect in that group receiving placebo. O'Nion et al[44] reported that zinc supplementation in a group of chronic hemodialysis patients appeared to improve hypogeusia and oral intake, suggesting a relationship between zinc metabolism and taste perception. It is worth noting, however, that other studies in patients with renal failure have failed to correlate taste acuity with zinc levels, pointing out that while hemodialysis seemed to improve hypogeusia, a similar effect on circulating zinc was not observed.[45]

Keiser et al[46] described a complete loss of taste and depleted zinc levels in a patient with cystinuria treated with penicillamine. Zinc therapy restored both to normal. Carrying this one step further, Wright and associates[47] created zinc depletion in six healthy adult males and studied their ability to taste. Subjects were given four concentrations of sodium chloride (salty) and urea (bitter) solutions. The only alteration observed was a decrease in the perceived intensity for the lower concentrations of the salty solution that seemed to improve with zinc replacement.

In 1971, Henkin et al[48] described a syndrome of idiopathic hypogeusia in 35 patients. Twenty-two of these patients also complained of dysgeusia, 28 complained of a smell loss, and 16 complained of dysosmia. Taste testing was based upon a forced-choice, three-stimulus drop technique, while olfactory testing was based upon a similar forced-choice, three-stimulus sniff technique.[48] Interestingly, test results indicated that all patients were both hypogeusic and hyposmic, albeit to varying degrees. These patients were treated empirically with zinc supplements, and this was said to correct many of their symptoms, but more specific information was not provided.

In a follow-up study, the same investigators studied the effect of zinc supplements in a group of 47 patients with hypogeusia.[49] In a single-blind fashion, patients were given placebo for 1 to 4 weeks, followed by a low or high dose of zinc. While the lower dose produced only a slight improvement in taste acuity, the higher dose restored over 50% of all taste thresholds for all of the taste qualities to the normal range. Despite the lack of a double-blind, randomized design and the absence of a control group, this study fueled the notion that zinc may be a viable, general treatment for taste loss.

The evidence linking zinc metabolism and olfactory loss has been even more tenuous. Kreuger and Kreuger[50] published a case report describing a 67-year-old patient with idiopathic hyposmia and dysosmia, a history of alcoholism, and a low serum zinc level. Zinc administration resulted in subjective improvement of the olfactory impairment, but no objective testing was performed. Henkin and colleagues[51] reported six patients with progressive systemic sclerosis developing a syndrome of anorexia, taste and smell loss, mental changes, and cerebellar dysfunction while on histidine therapy. Increased urinary zinc excretion with depletion of serum zinc concentration was observed. The syndrome was reversed within 24 hours after administration of zinc.

The most definitive study to evaluate the therapeutic potential of zinc for taste and smell disorders was conducted by Henkin et al.[52] They looked at 106 patients presenting to their clinic with taste or smell dysfunction secondary to a variety of etiologies, but not including any of the known zinc-wasting syndromes listed above. A randomized, double-blind crossover design was used to compare zinc sulfate with placebo. Interestingly, the mean serum zinc levels were significantly lower for the treatment group as compared to normal volunteers. Nevertheless, the results revealed that zinc sulfate was no better than placebo in the treatment of these disorders.

It would appear from these various studies that zinc depletion may occasionally alter taste and, to a lesser degree, smell perception. Such alterations may in turn show some response to zinc replacement. However, in the absence of a known condition that predisposes to zinc deficiency, zinc is not effective therapy. It should not be used as a blanket treatment for taste or smell loss, including those situations where the loss remains idiopathic.

REFERENCES

1. Schiffman SS: Taste and smell in disease. N Engl J Med 308:1337–1343, 1983.
2. Duncan HJ, Seiden AM, Paik SI, Smith DV: Differences among patients with smell impairment resulting from head trauma, nasal disease, or prior upper respiratory infection. Chem Senses 16:517, 1991.
3. Apter AJ, Mott AE, Cain WS, Spiro JD, Barwick MC: Olfactory loss and allergic rhinitis. J Allergy Clin Immunol 90:670–680, 1992.
4. Smith DV: Taste and smell dysfunction. In: Paparella MM, Shumrick DA, Gluckman JL, Meyerhoff AA (eds): Otolaryngology–Head and Neck Surgery, vol. 3. Philadelphia: WB Saunders, 1990.
5. Smith DV, Frank RA, Pensak ML, Seiden AM: Characteristics of chemosensory patients and a comparison of olfactory assessment procedures. Chem Senses 12:698, 1987.
6. Pryse-Phillips W: Disturbance in the sense of smell in psychiatric patients. Proc R Soc Med 68:472, 1975.
7. Leopold D: Distorted olfactory perception. In: Doty RL (ed): Handbook of Olfaction and Gustation. New York: Marcel Dekker, 1995.
8. Mattes RD: Nutritional implications of taste and smell disorders. In: Doty RL (ed): Handbook of Olfaction and Gustation. New York: Marcel Dekker, 1995.
9. Kveton JF, Bartoshuk LM: The effect of unilateral chorda tympani damage on taste. Laryngoscope 104:25–29, 1994.
10. Scott AE: Clinical characteristics of taste and smell disorders. Ear Nose Throat J 68:297, 1989.
11. Seiden AM: The diagnostic evaluation of conductive olfactory loss. Presented as a candidate's thesis to the American Rhinologic, Otologic, and Laryngologic Society, 1995.
12. Goodspeed RB, Gent JF, Catalanotto FA: Chemosensory dysfunction: Clinical evaluation results from a taste and smell clinic. Postgrad Med 81:251–260, 1987.
13. Deems DA, Doty RL, Settle RG, et al: Smell and taste disorders, a study of 750 patients from the University of Pennsylvania Smell and Taste Center. Arch Otolaryngol Head Neck Surg 117:519, 1991.
14. Costanzo R, Becker DP: Smell and taste disorders in head injury and neurosurgery patients. In: Meiselman HL, Rivlin RS (eds): Clinical Measurement of Taste and Smell. New York: Macmillan, 1986.
15. Douek E, Bannister LH, Dodson HC: Recent advances in the pathology of olfaction. Proc Soc Med 68:467–470, 1975.
16. Jafek BW, Hartman D, Eller PM, Johnson EW, Strahan RC, Moran DT: Postviral olfactory dysfunction. Am J Rhinol 4(3):91–100, 1990.
17. Duncan HJ, Seiden AM: Long-term follow-up of olfactory loss secondary to head trauma and upper respiratory tract infection. Arch Otolaryngol Head Neck Surg 121:1183–1187, 1995.
18. Seiden AM, Smith DV: Endoscopic intranasal surgery as an approach to restoring olfactory function. Chem Senses 13:736, 1988.
19. Amoore JE: Effects of chemical exposure on olfaction in humans. In: Barrow CS (ed): Toxicology of the Nasal Passages. Washington, DC: Hemisphere Publishing, 1986.
20. Hornung DE, Schwob JE: Congenital lack of olfactory ability. Ann Otol Rhinol Laryngol 101:229–236, 1992.

21. Paik SI, Seiden AM, Duncan HJ, Smith DV: Olfactory mucosal biopsy in patients with congenital anosmia. Chem Senses 16:566, 1991.
22. Murphy C: Taste and smell in the elderly. In: Meiselman HL, Rivlin RS (eds): Clinical Measurement of Taste and Smell. New York: Macmillan, 1986.
23. Paik S, Lehman MN, Seiden AM, Duncan HJ, Smith DV: Human olfactory biopsy. Arch Otolaryngol Head Neck Surg 118:731–738, 1992.
24. Bartoshuk LM, Gent J, Catalanotto FA, Goodspeed RB: Clinical evaluation of taste. Am J Otolaryngol 4:257–260, 1983.
25. Goodspeed RB, Catalanotto FA, Gent JF: Clinical characteristics of patients with taste and smell disorders. In: Meiselman HL, Rivlin RS (eds): Clinical Measurement of Taste and Smell. New York: Macmillan, 1986.
26. Seiden AM, Duncan HJ, Smith DV: Office management of taste and smell disorders. Otolaryngol Clin North Am 25:817–835, 1992.
27. Cain WS, Goodspeed RB, Gent JF, Leonard G: Evaluation of olfactory dysfunction in the Connecticut chemosensory clinical research center. Laryngoscope 98:83–88, 1988.
28. Leopold DA, Preti G, Mozell MM, Youngentob SL, Wright HN: Fish-odor syndrome presenting as dysosmia. Arch Otolaryngol Head Neck Surg 116:354–355, 1990.
29. Doty RL, Bartoshuk LM, Snow JB: Causes of olfactory and gustatory disorders. In: Getchell TV, Doty RL, Bartoshuk LM, Snow JB (eds): Smell and Taste in Health and Disease. New York: Raven Press, 1991.
30. Schiffman SS: Drugs influencing taste and smell perception. In: Getchell TV, Doty RL, Bartoshuk LM, Snow JB (eds): Smell and Taste in Health and Disease. New York: Raven Press, 1991.
31. Ophir D, Guterman A, Gross-Isseroff R: Changes in smell acuity induced by radiation exposure of the olfactory mucosa. Arch Otolaryngol Head Neck Surg 114:853–855, 1988.
32. Beidler LM, Smith JC: Effects of radiation therapy and drugs on cell turnover and taste. In: Getchell TV, Doty RL, Bartoshuk LM, Snow JB (eds): Smell and Taste in Health and Disease. New York: Raven Press, 1991.
33. Doty RL: Olfactory dysfunction in neurodegenerative disorders. In: Getchell TV, Doty RL, Bartoshuk LM, Snow JB (eds): Smell and Taste in Health and Disease. New York: Raven Press, 1991.
34. Kadi J, Greer RO, Jafek BW: Oral evaluation of patients with chemosensory disorders. Ear Nose Throat J 68:373–380, 1989.
35. Bartoshuk LM, Kveton J: Peripheral source of taste phantom (i.e dysgeusia) demonstrated by topical anesthesia. Chem Senses 16:499, 1991.
36. Doty RL, Shaman P, Dann M: Development of the University of Pennsylvania Smell Identification Test: A standardized microencapsulated test of olfactory function. Phys Behav 32:489–502, 1984.
37. Li C, Yousem DM, Doty RL, Kennedy DW: Neuroimaging in patients with olfactory dysfunction. Am J Rhinol 162:411–418, 1994.
38. Zinreich SJ, Kennedy DW, Rosenbaum AE, Gayler BW, Kumar AJ, Stammberger H: CT of nasal cavity and paranasal sinuses: Imaging requirements for functional endoscopic sinus surgery. J Radiol 163:769–775, 1987.
39. Russell RM, Cox ME, Solomons N: Zinc and the special senses. Ann Intern Med 99:227–239, 1983.
40. Lovell MA, Jafek BW, Moran DT, Rowley JC: Biopsy of human olfactory mucosa. Arch Otolaryngol 108:247–249, 1982.
41. Moran DT, Jafek BW, Rowley JC, Eller PM: Electron microscopy of olfactory epithelia in two patients with anosmia. Arch Otolaryngol 111:122–126, 1985.
42. Fera MAD, Mott AE, Frank ME: Iatrogenic causes of taste disturbances. In: Doty RL (ed): Handbook of Olfaction and Gustation. New York: Marcel Dekker, 1995.
43. Mahajan SK, Prasad AS, Lumbujon J, Abbasi AA, Briggs WA, McDonald FD: Improvement of uremic hypogeusia by zinc: A double-blind study. Am J Clin Nutr 33:1517–1521, 1978.
44. O'Nion J, Atkin-Thor E, Rothert SW: Effect of zinc supplementation on red blood cell zinc, serum, zinc taste acuity, and dietary intake in zinc deficient dialysis patients. Dialysis Transpl 7:1208–1213, 1978.
45. Vremon HJ, Venter C, Leegwater J, Oliver C, Weiner MW: Taste, smell, and zinc metabolism in patients with chronic renal failure. Nephron 26:163–170, 1980.
46. Keiser HR, Henkin RI, Bartter FC, Sjoerdsma A: Loss of taste during therapy with penicillmine. JAMA 203:381–383, 1968.
47. Wright AL, King JC, Baer MT, Citron LJ: Experimental zinc depletion and altered taste perception for NaCl in young adult males. Am J Clin Nutr 34:848–852, 1981.
48. Henkin RI, Schechter PJ, Hoye R, Mattern CF: Idiopathic hypogeusia with dysgeusia, hyposmia, and dysosmia. JAMA 217:434–440, 1971.
49. Schechter PJ, Friedewald WT, Bronzert DA, Raff MS, Henkin RI: Idiopathic hypogeusia: A

description of the syndrome and a single blind study with zinc sulfate. In: Pfeiffer C (ed): International Review of Neurobiology. New York: Academic Press, pp 125–140, 1972.

50. Kreuger KC, Kreuger WB: Hypogeusia and hyposmia associated with low serum zinc levels, a case report. J Am Med Women Assoc 35:109–111, 1980.

51. Henkin RI, Patten BM, Re PK, Bronzert DA: A syndrome of acute zinc loss. Arch Neurol 32:745–751, 1975.

52. Henkin RI, Schecter PJ, Friedewald WT, Demets DL, Raff M: A double blind study of the effects of zinc sulfate on taste and smell dysfunction. Am J Med Sci 272:285–299, 1976.

3

Basic Anatomy and Physiology of Olfaction

WILLIAM T. NICKELL, Ph.D.

The human olfactory epithelium is a 1-cm^2 patch sequestered deep in each nasal cavity. Odorous substances may reach the epithelium by way of normal respiration through the nose or by a back route through the nasal pharynx. On reaching the epithelium, odorant molecules must dissolve in an aqueous medium, the mucus covering the epithelium, before interacting with membrane-bound receptors on primary olfactory neurons (PONs). These small bipolar neurons, uniquely, reside outside the central nervous system in the epithelium. Each PON sends a single unbranched dendrite to the surface of the epithelium. A number of cilia extend from the tip of each dendrite into the layer of mucus covering the epithelium. Protein odor receptors are localized on these cilia.

Primary olfactory neurons must detect the presence of odors and code their responses in a way that allows the central nervous system to identify these odors over a wide range of concentrations. Although olfactory research has traditionally lagged behind some other sensory systems, work during the last decade has advanced understanding of olfactory transduction mechanisms to a level comparable to that of vision and hearing. The mucus covering the epithelium has been shown to contain unique odorant-binding proteins that may aid in dissolving hydrophobic odorants in the aqueous mucus. Odors are detected by protein receptors homologous to the neural and hormonal receptors that are coupled to second messengers. Binding of odorants to these receptors is coupled to spike-generating mechanisms of the olfactory neurons through the cyclic adenosine monophosphate second-messenger system. In contrast to color vision and taste, where three primary colors or four primary tastes are used to code a very large number of sensations, it now appears that there are many hundreds of olfactory receptor proteins each responding to a different spectrum of odors.

The vertebrate olfactory system can detect and identify nearly any substance that is sufficiently volatile to produce a measurable vapor pressure, including newly synthesized molecules not present in nature. To do this, olfactory receptors must respond to an extraordinarily large variety of chemical substances. In addition, each molecule must produce a unique pattern of neural responses that the central nervous system (CNS) can use to distinguish among odorants. Until recently, the transduction and coding mechanisms that accomplish these tasks were not known. Within the last decade, however, a series of important investigations have raised our understanding of olfactory transduction and coding to a point comparable to that of other sensory systems. In this chapter I will describe our current knowledge of olfactory transduction and coding, in the context of earlier theories and speculations.

ANATOMY OF THE OLFACTORY EPITHELIUM

Location of the Olfactory Epithelium

The receptive surface for the human olfactory system is an approximately 1-cm^2 patch of specialized epithelium located within each nasal vault out of the most direct path of air flow during normal respiration. This dimension may be compared with an area of approximately 4.5 cm^2 in the rabbit.[1] Because of its location, the concentration of odorants at the epithelium can be partially controlled by voluntary respiratory movements. The receptive surface is also accessible by a backward path through the nasal pharynx. This path is important for perception of flavor, as it allows the odors of food in the mouth to reach the olfactory receptors during chewing.

The Primary Olfactory Neuron

The receptive cell of the olfactory system is the primary olfactory neuron (PON; Fig. 3–1). The cell bodies of these neurons are positioned in the olfactory epithelium, far from the CNS and exposed to harmful agents that may be brought in through the nose. This vulnerability may account for the unique ability of PONs to regenerate from basal cells, even in adults.[2] Primary olfactory neurons are small bipolar neurons; the cell body is typically 5 to 8 μm in diameter. Primary olfactory neuron soma are positioned in the inner two thirds of the epithelium. Each soma sends a single 1- to 2-μm diameter dendrite to the surface of the epithelium; the length of this dendrite is thus determined by the position of the soma within the epithelium. The dendrite terminates in an *olfactory knob* near the surface of the epithelium. The olfactory knob supports a number (10 to 20) of cilia. These cilia exhibit the typical "9 + 2" arrangement of tubules found in motile cilia; however, microtubules in the olfactory cilia of mammals, including humans, are incomplete and immotile. The olfactory cilia in fish, amphibia, and reptiles are motile. Cilia are 0.3 μm at their base and taper to 0.1 μm at their tip. The motile cilia of amphibia are 200 μm in length; mammalian cilia are shorter (50 μm).

The axons of PONs are among the smallest (0.1 to 0.3 μm diameter) in the vertebrate nervous system and are unmyelinated. Because of their small diameter

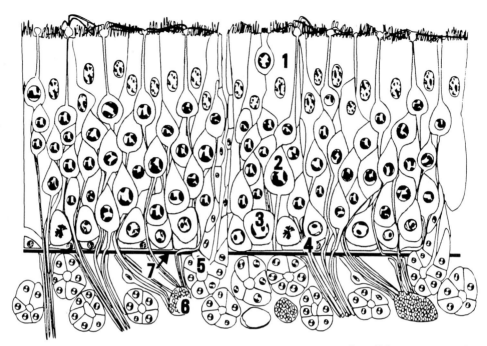

Figure 3–1.   Schematic diagram of the olfactory epithelium. 1 = sustentacular cell, 2 = mature receptor neuron, 3 = globose basal cell, 4 = pyramidal basal cell, 5 = Bowman's gland, 6 = olfactory nerve bundle, 7 = basement membrane.

and the absence of myelin, the conduction velocity is very low ($0.1\,\text{m/s}$[3,4]). Axons are grouped into fascicles of 50 to 100 axons surrounded by a Schwann cell sheath. This arrangement is unusual and typical of developing neural connections rather than a mature nerve. Because the diameter of the axons is smaller than the resolution of the optical microscope, these fascicles were initially mistaken for single fibers, leading to the false conclusion that olfactory axons branched within their target, the olfactory bulb.[5]

Axons of PONs synapse with second-order olfactory neurons (mitral and tufted cells) in the olfactory bulb; the transmitter at this synapse has only recently been identified. The early candidate as transmitter was carnosine, a dipeptide (β-alanyl-L-histidine); the synthetic machinery for carnosine is present in PON neurons and terminals[6,7] and binding sites are present,[8] but clear physiological actions of carnosine could not be demonstrated.[9,10]

Recent studies show that the transmitter is an excitatory amino acid (EAA). Histological studies demonstrate the presence of synthetic enzymes for glutamic acid in the terminals of the olfactory nerve,[11] and in the isolated turtle olfactory bulb the intracellularly recorded response to olfactory nerve stimulation is blocked by EAA antagonists and the response is partially mimicked by exogenous application of EAAs. Both N-methyl-D-aspartate (NMDA) and non-NMDA receptors appear to be present.[3]

Structure of the Epithelium

Primary olfactory neuron cell bodies are tightly packed within the basal two thirds of the epithelium (Fig. 3–1), resulting in a high density of sensory neurons. As a result of this high density of PONs, their cilia form a dense mat in the mucous layer that covers the epithelium. In mammals, the number of ciliated dendritic endings on the epithelial surface of mammals vary from 5 to 43×10^4 per mm^2, with most reports in the range of 10 to 20×10^4 per mm^2 (reviewed by Farbman[12]).

Near the epithelial surface, PON dendrites are surrounded by the cell bodies and nuclei of *supporting cells*. Each supporting cell sends a thin process that winds through the PONs to the basement membrane, while the apical surface of the supporting cell is covered with microvilli. Tight junctions between the dendrites and supporting cells form a barrier to flow of both ions and molecules between the mucus and the epithelium. Besides mechanically isolating PON cell bodies from the environment, at least two metabolic functions are attributed to supporting cells. First, microvilli, together with large quantities of smooth endoplasmic reticulum and intramembranous rod-shaped particles found within the supporting cells, are consistent with a role in regulation of fluid and ion flow between the epithelium and the mucous layer immediately above it. This regulatory activity is critical because the transduction functions of the cilia presumably require close regulation of their ionic environment. Second, these cells contain very high levels of the cytochrome P-450 enzyme system,[13] which is specialized for degradation of several classes of organic molecules. In the epithelium, this system may serve to degrade olfactory stimulants or to remove potentially toxic substances.

Bowman's glands are primarily responsible for production of the mucus that covers the epithelium. Bowman's glands are located in the lamina propria beneath the epithelium and communicate with the surface by a duct that passes through the epithelium. Secretion from Bowman's glands appears to be stimulated by both sympathetic and parasympathetic innervation.[14,15]

PERIRECEPTOR EVENTS

The mucus that covers the olfactory epithelium plays important roles in olfactory transduction (Fig. 3–2). These roles are designated "perireceptor events."[16] A first and obvious role of the mucus is maintenance of a proper ionic environment for the neural functions of the cilia and the PONs. Reliable operation of the transduction process requires proper osmotic pressure, maintenance of an intracellular potential in the PON and cilia, ions necessary to carry the generator currents that depolarize the PON, and initiation of action potentials that can propagate to the brain. The minimum set of ions that must be regulated to maintain these functions are Na$^+$ (action potential and generator currents), K$^+$ (to regulate the resting potential), and Ca^{++} (necessary for numerous membrane functions). Chloride is necessary to balance the cations. The maintenance of this ionic environment after secretion of the mucus by Bowman's glands appears to be the responsibility of the supporting cells.

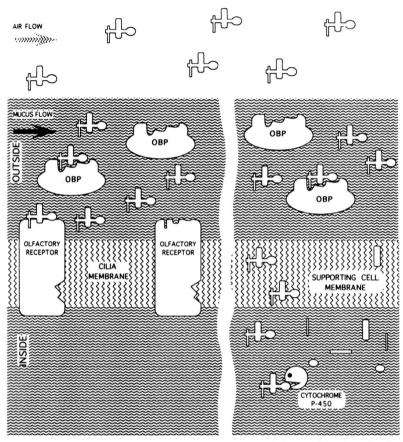

Figure 3–2.   Perireceptor events in olfactory transduction. Step 1: Odorant molecules in inspired air are dissolved in the aqueous mucus layer above the epithelium. Transfer of hydrophobic molecules to the mucus may be aided by odorant binding protein (OBP). Step 2: Odorant molecules bind to protein receptors embedded in the membrane of the cilia of PONs. Step 3: Removal of odorants. Odorant molecules may be carried away by the flow of mucus, by binding to OBP, or by passing through the supporting cell membrane. Inside the supporting cell the odorant may be destroyed by degradative enzymes or may pass to the tissue space below the epithelium where it would be carried away by the circulation.

A second function of the mucus is the transfer of odorants from the air into the aqueous environment of the cilia. Odorants are volatile, often hydrophobic, compounds. To interact with receptors on the cilia of PONs, these molecules must be transferred from air to the aqueous medium surrounding the cilia. The efficiency of this transfer can be estimated from the air-water partition coefficient of the odorant. Surprisingly, such calculations show that at equilibrium even hydrophobic odorants are slightly more concentrated in water than in air.[17,18] Thus, the transfer of molecules from air to mucus provides a first mechanism for concentration or amplification of the odor signal; increased solubility of hydrophobic compounds in mucus would provide further concentration.

At least one protein present in mucus and having high affinity for hydrophobic odorants has been identified.[19,20] Odorant binding protein (OBP), secreted by Bowman's glands, is a small (18 kd) soluble protein; its amino acid sequence is homologous to proteins known to function as carriers of small hydrophobic molecules.[21] Odorant binding protein constitutes about 1% of the protein of the mucus covering the olfactory epithelium; its affinity for odorants is in the micromolar range.[22,23] The localization of OBP in the mucus and its homology with carrier molecules in other systems suggest that it may increase the solubility of hydrophobic odorants. Unfortunately, there is no direct evidence that OBP plays the postulated role as a carrier protein: there appear to be no measurements of the solubility of odorants in mucus and no theoretical calculations of the effectiveness of the protein as a carrier.

Olfactory binding protein was serendipitously identified during the search for the olfactory receptor. On the presumption that receptors would bind particularly potent odorants with high affinity, several laboratories attempted to isolate olfactory receptor proteins by biochemically isolating odorant binding fractions. This strategy was unsuccessful at finding olfactory receptors: the affinity of receptors for odorants is lower than the affinity of neural and hormonal receptors for drugs. In addition, there now appear to be a very large number of receptor types (which dilute the binding of any particular odorant). However, the search for a protein in olfactory epithelium that binds the odorant pyrazine with high affinity produced OBP. Olfactory binding protein was initially believed to be the long-sought olfactory receptor. Localization studies, however, demonstrated that the protein was located in the lateral nasal glands and Bowman's glands, rather than PONs.[24]

For continued function, odorants must also be removed from the vicinity of the receptors. As there is some concentration of odorants from air to mucus, removal of odorants by transfer back into air is ineffective. Flow of mucus away from the olfactory epithelium provides one mechanism for removal. Odorants may also diffuse into the large tissue space deeper than the epithelium, where they would be carried away in the circulation.[18] One suggested role for OBP is removal of odorants by binding and holding them away from receptor sites; as with other suggested roles for OBP, there is no direct empirical or theoretical evidence for this function. The degradative enzyme systems present in supporting cells may also remove odors, although their degradative action may produce new odorants.[18]

Primary olfactory neurons, because they are in contact with respired air and send axons directly to the brain, provide a potential route for transfer of pathogens or toxins to the brain. Such transfer has been demonstrated in animal models.[25] Thus, the presence of immune functions in the mucous layer and epithelium are of great importance. Olfactory mucus is similar to the mucus of respiratory epithelium in containing components of the secretory immune system. Bowman's glands secrete immunoglobulin A, responsible for immobilizing pathogens in mucus, as well as antimicrobial proteins lactoferrin and lysozyme.[26] The various metabolic enzymes present in the nasal cavity, including cytochrome P-450,[13] may also play a role in destruction of toxins that might otherwise injure the peripheral olfactory receptors or even the brain.[27,28]

PHYSIOLOGICAL RESPONSES OF PONS TO ODORANTS

The relative inaccessibility of the mammalian olfactory epithelium and the small size of mammalian PONs has limited most electrophysiological studies to amphibian preparations. Initial physiological studies of olfactory responses used the electro-olfactogram (EOG), which is a slow negative-going potential that can be recorded between the epithelial surface and deep layers during exposure to an odorant.[29] The EOG response was used to investigate the time course of the olfactory response,[30] the sensitivity of the receptors, and the topographic localization of sensitivity to odorants.[31–34] Rapid application of odorant to the mucous surface causes, after a delay of several hundred milliseconds, a negative-going potential. With continued application of the odorant, this negative potential rises to a peak in about 200 milliseconds and then declines to a plateau that may be maintained as long as the stimulus is applied (Fig. 3–3).

The EOG is produced by generator currents passing across the tight junctions between the receptors located on cilia in the mucus and PON somata in the epithelium. The delay between stimulus application and the initial response is the sum of the time required for stimulus molecules to diffuse across the mucous layer plus any delay created by the transduction mechanism. Calculations show that the time required for diffusion of odorants through the mucus is much shorter than the time between application of the odorant and the beginning of the EOG response. Thus, EOG recordings suggest that the olfactory transduction mechanism involves at least one slow step.

Electro-olfactogram recordings from multiple sites on the salamander ventral olfactory epithelium produce odorant-specific response patterns.[32] Localization of metabolic activity generated by odorants using 2-deoxyglucose also demonstrated odor-specific patterns on salamander and mouse epithelium.[35] The physiological significance of these patterns is questionable, however, as only a crude topographic organization is maintained in the projection from the epithelium to the olfactory bulb.[36–42]

More refined examination of the responses of PONs to odors requires recording from individual receptor neurons. Extracellular recordings of spike activity of single PONs became possible with the development of low-resistance metal-filled microelectrodes.[43] These early studies demonstrated that most PONs respond to a broad range of odorants, but that each PON has a characteristic response profile. However, later studies were unable to establish categories of receptors corresponding to identifiable chemical properties.[44,45] Quantitative studies of PON responses as a function of odor concentration show a remarkably small dynamic range: for most PONs the transition from no response to maximal response covers only 1 log unit of concentration.[45–51]

Direct investigation of olfactory transduction mechanisms required intracellular recordings. Because of their small size, impalement of PONs with a conventional sharp microelectrode proved very difficult.[45,52,53] Development of the "patch clamp" recording technique and of methods of dissociating PONs without use of proteolytic enzymes were required for successful studies of the biophysics of olfactory transduction. In patch clamp recording a polished pipette tip of relatively large

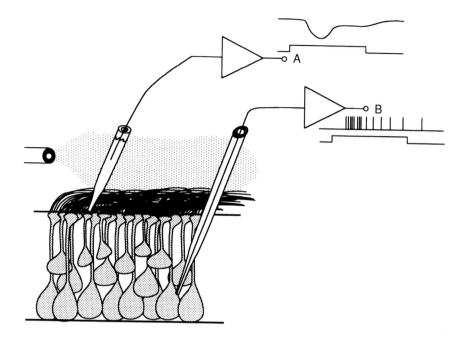

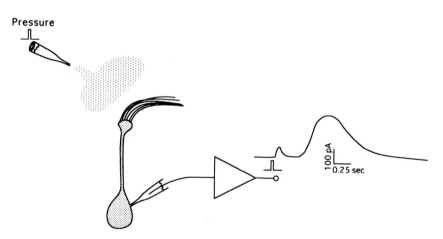

Figure 3–3.　Physiological recordings of responses of primary olfactory neurons to odors. Upper panel: electro-olfactogram and extracellular single-unit recording. **A**: A recording electrode placed on the surface of the epithelium records a slow negative-going potential in response to application of odorant to the epithelium. **B**: A smaller electrode pushed into the epithelium records action potentials from a single primary olfactory receptor neuron (PON). Lower panel: patch clamp recording from single PON. Patch clamp techniques provide a direct low-resistance connection to the inside of a single PON. Pressure ejection of a short puff of odorant cocktail containing 100 mmol/L KCl produces two inward (depolarizing) currents: a rapid depolarization caused by arrival of the KCl-containing solution followed by a slow current produced by olfactory transduction mechanisms.

(1 µm) diameter is pressed against the cell membrane where it forms a very tight seal (gigaseal). A variety of maneuvers are then used to rupture the membrane patch inside the pipette, allowing a direct low-resistance electrical connection with the interior of the cell. Electronics connected to the pipette can record either membrane voltage or, by holding the membrane potential constant, membrane currents. Alternatively, the patch may be left intact, allowing individual ion channels within the patch to be recorded.[54–56] As with other neurons, the membrane resistance of PONs measured with patch clamp recording is substantially higher than that observed with sharp electrodes; the membrane resistance of PONs is measured as 2 to 5 gigΩ (2 to 5 × $10^9 \Omega^{57–59}$). The practical significance of this high membrane resistance is that very small currents, such as might be generated by a single odor-activated channel, can depolarize the PON to threshold and produce an action potential.

Successful recordings of odorant-induced potentials and generator currents were first achieved by Firestein and Werblin[60] using both a slice preparation and mechanically dissociated salamander PONs. Odorants were applied to the cilia of the PON by pressure ejection from a pipette. This pipette contained, in addition to a "cocktail" of five odorants (acetophenone, amyl acetate, cineole, phenylethylamine, and triethylamine), an increased concentration (100 mmol/L) of K^+. Upon reaching the cilia, these K^+ ions produced immediate membrane depolarization by alteration of ionic equilibrium potentials. Thus, addition of the K^+ to the odorant cocktail allowed Firestein and Werblin to determine the exact time of arrival of the odorants at the cilia and, by calculation, to estimate their actual concentration.

The response of the PON membrane potential to the release of a bolus of this cocktail near the cilia of a PON is illustrated in Figure 3–3.[60,61] After a short delay, due to the time required for movement of the bolus from the pipette tip to the cilia, there is a transient inward (depolarizing) current. This current is due entirely to the increase in K^+ in the vicinity of the cilia; it allows estimation of the effective arrival time and odorant concentration at the cilia. There follows with a delay of several hundred milliseconds (minimum 140, mean 320 milliseconds) a second inward current.

This arrangement allowed the concentration of odorants at the cell to be adjusted by changing the width of the pressure pulse that ejected the cocktail. The dose-response relation for the generator current was sigmoid with a Hill coefficient of 2.7 and a $K_{1/2}$ (concentration of odorant at which the response was half maximal) of 28 µmol/L. The lowest estimated concentration at which a response could be detected was 6 µmol/L. The sigmoid dose-response curve indicates that the response is cooperative: at low concentration of odorant, a small increase in concentration produces a more-than-linear increase in the response. The estimated half-maximal and minimum detectable concentrations are higher than has been estimated from behavioral estimates of sensitivity of the olfactory system to odorants. The difference may arise because of reduced sensitivity of the PON in the experimental preparation or inaccurate estimates of the concentration of odorants in vivo. It is also possible that the PONs tested were not highly sensitive to any of the odorants used in the experiment. This points out the difficulties encountered when performing such experiments.

The current-voltage curve for the generator conductance is linear, with a reversal potential of about $+5\,mV$, suggesting a nonspecific cation conductance. This finding suggests that the transduction mechanism opens a simple pore in the PON membrane, through which any monovalent positive ion can pass. If the current involved, for example, a carrier protein, the current capacity would be limited and the relation between current and voltage would not be a straight line; if the channel were selective for Na^+, for example, the reversal potential would be near the $+20\,mV$ reversal potential of Na^+.

For longer periods of stimulus application, up to 1 second, the response amplitude is more closely proportional to the product of concentration $\times$ time rather than to the maximum concentration of odorant; with constant stimulus application, the response declines after about 4 seconds.[61] The observation that the odor response is proportional to the product of concentration and time suggests that the PON is able to "integrate" the stimulus produced by low concentrations of odorant, increasing the sensitivity of the olfactory system to weak signals. The decline in response after several seconds exposure to the stimulus is consistent with previous findings with extracellular single unit and EOG recordings. In this respect the olfactory system is similar to other sensory systems in exhibiting fatigue or adaptation to prolonged application of stimulus.

TRANSDUCTION MECHANISMS IN CILIA

The two fundamental questions about any sensory system are transduction and coding: how does the stimulus produce neural activity and how is that activity patterned so that the nervous system can interpret it? Understanding of these issues in olfaction lagged after other sensory systems for several reasons: (1) the olfactory epithelium is relatively inaccessible and physical control of stimulus strength and timing is difficult, (2) the neurons are small and difficult to impale with a sharp electrode, (3) the physical dimensions of the stimulus are not obvious, and, of course, (4) the funding and interest in olfaction have always been lower than for sensory systems more impressive to visually oriented primates. Nevertheless, research in numerous laboratories over the past decade has brought our understanding of olfactory transduction and coding to a level comparable to that of other sensory systems.

In the absence of real data, several hypotheses, some quite imaginative, have been maintained energetically as the mechanisms of transduction. The simplest proposed transduction mechanism notes that numerous odorants will alter the electrical properties of artificial membranes and cultured nonolfactory neurons. Cultured neuroblastoma cells respond to some odorants in the same concentration range as olfactory neurons, and the response properties of the cultured cells can be altered by changing the lipid concentrations of their membranes. Thus, olfactory receptor neuron membranes might differ in their composition sufficiently to provide a capacity for discrimination. Although this hypothesis is still maintained by some workers,[62] most evidence supports the existence of some kind of specific protein receptor.

The evidence for specific olfactory receptor proteins falls into several categories:

1. *Specific anosmias.* Certain individuals are unable to detect specific odorants; this deficit is sometimes transmitted according to Mendelian genetics.[63,64] Presumably, the deficit is due to loss of a specific protein receptor.
2. *Stereospecificity of odorants.* In some stereo pairs of odorants the two optical isomers have entirely different odor quality.[65] This argues strongly for specific binding to a selective site.
3. *Saturable binding.* There is evidence of reversible and saturable binding in some olfactory tissues,[66] although it is difficult to demonstrate that this binding results from saturation of olfactory receptors and not from other receptors within the transport system, including degradative enzymes or odorant binding protein.

Several classes of proteins have been considered as candidate olfactory receptors. The ligand-gated membrane channels were initially considered plausible models for olfactory receptors, in analogy with the best-understood neural receptors at the time—the nicotinic acetylcholine receptor. Such channels would depolarize PONs through direct coupling between an odorant binding site and an ion channel. The immune system, which also has the capacity to identify a nearly limitless number of substrates, was also suggested as a model of olfactory receptors,[18] although the method of coupling to membrane depolarization was never clarified. A consensus has developed, however, that olfactory receptors are of the class of protein receptors that pass seven times through the plasma membrane and couple to second-messenger systems within the cell. Such protein receptors bear a family resemblance to the receptors to numerous hormones and neurotransmitters; this familial resemblance allowed rapid progress once the receptor type was identified.

Several lines of evidence point to second messenger–coupled receptors as the transduction medium for olfactory reception. First, biochemical studies showed that the components of a G protein–coupled system are present in olfactory tissue. Olfactory tissues contain specific adenylate cyclases that can be stimulated by odorants[23,67–70] and cation selective channels that are opened by cyclic adenosine monophosphate (cAMP).[71–76] There is solid evidence that the inositol triphosphate pathway is involved in olfactory transduction in arthropods,[77–79] but the role of this pathway in vertebrate olfaction is not clear.

Second, when direct recording of odorant responses from isolated PONs was accomplished,[60] characteristics of the response were consistent with the characteristics expected of a second messenger–coupled system but not of a ligand-gated or nonspecific system. Following rapid application of a cocktail of odorants to the cilia of an isolated PON, there is a delay of several hundred milliseconds before a depolarization is observed. This delay is presumably accounted for by the time necessary for generation of second messenger.

Initial attempts to isolate the receptor sought to isolate proteins binding potent odorants. These studies led to the identification of OBP, but not to the olfactory receptor. Presumably, the odorants bind to their receptors too weakly or the number of receptor types is too large to allow separation of receptor-odorant binding from other kinds of binding that are bound to occur in olfactory tissue. In the end, isolation of the olfactory receptor has resulted from the developing consensus that the olfactory receptor was a G protein–coupled receptor. Aided by improved molecular genetic techniques, and based on suspected sequence homologies with other seven-membrane domain receptors, a family of several hundred to a thousand distinct genes that are unique to PONs has been isolated.[80] These genes are now generally conceded to code for the olfactory receptor.[81–83]

The steps in olfactory transduction involving the cAMP system are illustrated in Figure 3–4. An odorant molecule dissolved in the aqueous mucus or carried on an OBP becomes bound to a seven-transmembrane region receptor. The odorant-receptor complex possesses catalytic activity that converts a membrane soluble protein designated G_s into an active form. G_s (for stimulatory guanosine triphosphate [GTP] binding) normally diffuses in the plane of the membrane in conjunction with a bound molecule of guanosine diphosphate. After activation by the enzymatic activity of the odorant-receptor complex, the active portion of the G protein binds a molecule of GTP; this complex diffuses in the plane of the membrane until it binds to and activates an adenylate cyclase (AC) unit. The activated AC enzyme proceeds to convert cytoplasmic adenosine triphosphate to cyclic AMP, thus earning its name. Within olfactory cilia are Na^+-K^+ channels that are opened by cAMP.[74,84,85] Opening of these channels completes the process of transduction: incoming Na^+ depolarizes the PON via the cilium, resulting in cell depolarization and action potentials. The inward Na^+ current is presumably the generator current that is the cause of the potential measured in the EOG.

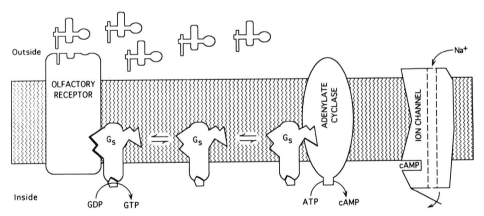

Figure 3–4. Steps in olfactory transduction. (1) Odorant molecule binds to an olfactory receptor protein. (2) Catalytic activity of the odorant-receptor complex converts a membrane-soluble protein designated G_s into an active form. (3) The active portion of the G protein binds a molecule of guanosine triphosphate (GTP); this complex diffuses in the plane of the membrane until it binds to and activates an adenylate cyclase (AC) unit. (4) The activated AC enzyme proceeds to convert cytoplasmic adenosine triphosphate (ATP) to cyclic adenosine monophosphate (cAMP). (5) cAMP binds to a membrane cation channel, opening it and allowing influx of Na^+, which depolarizes the PON and results in firing of action potentials.

This model, based on the work of many laboratories, clarifies old mysteries of olfactory transduction. Olfactory transduction does not require unique or exotic hypotheses; it is accomplished by mechanisms already familiar in other sensory systems and in neuronal and hormonal signaling. The rather complex cascade of events required by the cAMP signaling pathway provides advantages for transduction of olfactory signals. The process provides several stages of amplification, which provide increased sensitivity, but also accounts for the limited dynamic range of olfactory neurons.

Thus, as the result of work in numerous laboratories, the mechanism of transduction of the olfactory signal appears to be well understood. The method of coding of olfactory quality, however, was not determined by these studies. Additional studies using the techniques of molecular biology were required to provide a better understanding of olfactory coding.

PERIPHERAL CODING OF OLFACTORY QUALITY

The sensory qualities of taste and color vision can be broken down into primary qualities that reflect the properties of the peripheral receptors. In analogy, a substantial literature has attempted to identify primary olfactory qualities and to correlate these with physical or chemical properties of the odorant molecules. Success in this endeavor, it was hoped, would lead to insights into the mechanisms of olfactory transduction and provide a physical basis for perceived odor quality. Proposed determinants of olfactory quality included Raman infrared spectra (indicating low-energy molecular vibrations), functional groupings, positions of functional groups, and overall molecule shape (reviewed by Schiffmann[86,87]). Schiffman[86,87] used multidimensional scaling techniques to demonstrate that, while combinations of physical qualities could predict certain odor qualities, none of the theories adequately predicted odor quality from molecular structure. A very different hypothesis, suggested by Mozell and colleagues,[88–91] is that the olfactory epithelium functions as a gas chromatograph and that odor quality is determined by spatial and temporal patterns created on the epithelium as air passes over it. This hypothesis would place little emphasis on the discriminatory capabilities of the olfactory receptors; discrimination of odor quality would be shifted to central processing.

Using extracellular microelectrode techniques, numerous workers, starting with Gesteland and coworkers,[30,43] attempted to find classes of olfactory receptors according to the responses of single PONs to batteries of odorants. These studies generally agree that amphibian PONs are broadly tuned; each PON responds to about half of the odorants presented to it. Some odorants stimulate nearly all PONs, while others stimulate a smaller fraction. These studies did not, however, identify classes of PONs responding to identifiable chemical groupings. Instead, each PON tested appeared to have a unique profile of responses to the battery of odorants used.[92]

The coding theory that has received the most attention is that of Amoore and collaborators.[93–95] Amoore initially proposed a set of seven "primary odors"—camphoraceous, musky, floral, pepperminty, ethereal, pungent, and putrid. According to this theory the first five of these qualities were determined by the shape of the molecule. For example, spherical molecules of 0.7 nm diameter have a camphoraceous odor. Pungent and putrid odors are determined by the electronegativity of the molecule.

The attractiveness of this theory is obvious. If confirmed, it would provide a mechanism for predicting the odor quality of chemicals, of synthesizing odors of any quality, or of masking unpleasant odors. Unfortunately, although it received wide attention, the theory never satisfactorily accounted for the olfactory quality of many molecules. The authors continued to elaborate on this theory, using the occurrence of "specific anosmias" as indicators of primary odors. Such congenital anosmias presumably represent the absence of specific receptors in analogy with color blindness. Unfortunately for the stereochemical theory, the number of specific anosmias and thus of primary odors has now grown to nearly 50,[95] a number that renders the concept of "primary odor" more or less useless.

As in many other areas of biology, the techniques of molecular biology have clarified issues that defied the powers of less advanced techniques. As described above, using molecular genetics and the developing certainty that olfactory receptors are G protein–coupled receptors, Buck and Axel[80] were able to identify a family of genes homologous to the β-adrenergic receptor. This family consists of several hundred to a thousand genes grouped into several subfamilies. Subsequent work has shown that the genes are stable (do not undergo random recombination in the manner of the immune system), that each PON probably expresses only one olfactory receptor, that individual families are localized to broad areas of the epithelium, and that PONs expressing each gene project to a single glomerulus in the olfactory bulb.[81,82,83,96,97]

These findings, particularly the unexpectedly large number of receptor classes, clarify several questions raised by previous research. Attempts to classify PONs were unsuccessful because, with so many receptor types, each PON isolated by a microelectrode was unique; the probability of recording from even two PONs of the same class in a microelectrode study is quite small. Likewise, attempts to identify primary odors were unsuccessful because the number of receptor types is far too large for the available techniques. The 50 or so identified specific anosmias are likewise understandable as a small fraction of the number of genes actually present.

The current understanding, then, is that while olfactory transduction uses mechanisms very similar to those of other sensory systems, olfactory coding differs in fundamental ways from taste and color vision. Instead of using a small number of primary qualities to construct a very large number of possible sensations, the olfactory system uses a very large number of receptor classes to detect an even larger number of distinct odors. The large number of receptors raises the theoretical possibility that each odor stimulates one, or a very small number, of receptors; however, as electrophysiological studies are unanimous in finding that PONs typically respond to a substantial fraction of odors, it appears that olfactory quality depends on the *pattern* of activity across the population of receptors. Also, unlike some other sensory systems, topographic organization does not appear to play a large part in the processing of olfactory information. While there is some preservation of topography in the projection from the epithelium to the olfactory bulb, the real pattern is apparently the segregation of fibers from the distinct classes of

PONs into individual glomeruli in the olfactory bulb. This accounts for the patterning of metabolic activity in the olfactory bulb following exposure to particular odorants.

It is important to note the things *not* explained by these important findings. The gene products of these genes have not been tested for odor specificities. Given the large number of receptor classes, this task represents an enormous labor that may never be accomplished. Each receptor must be expressed in a model system and then tested against a battery of odorants. Without these data we do not have answers to the questions that motivated much of the earlier research in olfaction. Are psychological odor categories—pleasant, unpleasant, musky, pungent, etc—determined peripherally by the receptors, or are these qualities determined by central mechanisms? Especially given the current interest in olfaction as a model system for understanding of memory and central processing, these questions remain of great interest.

Acknowledgments—Dr. Marian Miller produced the excellent drawing of the olfactory epithelium (Fig. 3–1).

REFERENCES

1. Allison AC, Warwick RTT: Quantitative observations on the olfactory system of the rabbit. Brain 72:186–197, 1949.
2. Graziadei PPC, Monti-Graziadei GA: Continuous nerve cell renewal in the olfactory system. In: Jacobson M (ed): Development of Sensory Systems. New York: Springer-Verlag, 1978.
3. Berkowicz DA, Trombley PQ, Shepherd GM: Evidence for glutamate as the olfactory receptor cell neurotransmitter. J Neurophysiol 71:2557–2561, 1994.
4. Freeman WJ: Spatial divergence and temporal dispersion in the primary olfactory nerve of cat. J Neurophysiol 35:733–744, 1972.
5. Pinching AJ, Powell TPS: The neuropil of the glomeruli of the olfactory bulb. J Cell Sci 9:347–377, 1971.
6. Margolis FL: Carnosine. Trends Neurosci 1:42–44, 1978.
7. Rochel S, Margolis FL: Carnosine release from olfactory bulb synaptosomes is calcium-dependent and depolarization-stimulated. Brain Res 202:373–386, 1980.
8. Hirsch JD, Grillo M, Margolis FL: Ligand binding studies in the mouse olfactory bulb: Identification and characterization of a l(^{3}H) carnosine binding site. Brain Res 158:407–422, 1978.
9. Macleod NK, Straughan DW: Responses of olfactory bulb neurones to the dipeptide carnosine. Exp Brain Res 34:183–188, 1979.
10. Gonzalez-Estrada MT, Freeman WJ: Effects of carnosine on olfactory bulb EEG, evoked potentials and DC potentials. Brain Res 202:373–386, 1980.
11. Sassoe-Pognetto M, Cantino D, Panzanelli P, et al: Presynaptic co-localization of carnosine and glutamate in olfactory neurons. NeuroReport 5:7–10, 1993.
12. Farbman AI: Cell Biology of Olfaction. Cambridge, UK: Cambridge University Press, 1992.
13. Nef P, Heldman J, Lazard D, et al: Olfactory-specific cytochrome P-450. cDNA cloning of a novel neuroepithelial enzyme possibly involved in chemoreception. J Biol Chem 264:6780–6785, 1989.
14. Getchell ML, Getchell TV: Fine structural aspects of secretion and extrinsic innervation in the olfactory mucosa. Microscopy Res Tech 23:111–127, 1992.
15. Zielinski BS, Getchell ML, Wenokur RL, Getchell TV: Ultrastructural localization and identification of adrenergic and cholinergic nerve terminals in the olfactory mucosa. Anat Rec 225:232–245, 1989.
16. Getchell TV, Margolis FL, Getchell ML: Perireceptor and receptor events in vertebrate olfaction. Prog Neurobiol 23:317–345, 1984.
17. Senf W, Menco BPH, Punter PH, Duyventeyn P: Determination of odour affinities on the dose-response relationships of the frog's electro-olfactogram. Experientia 36:213–215, 1980.
18. Lancet D: Vertebrate olfactory reception. Ann Rev Neurosci 9:329–355, 1986.

19. Bignetti E, Cavaggioni A, Pelosi P, Persaud KC, Sorbi RT, Tirindelli R: Purification and characterization of an odorant-binding protein from cow nasal tissue. Eur J Biochem 149:227–231, 1985.

20. Pevsner J, Trifilette RR, Strittmatter SM, Snyder SH: Isolation and characterization of an olfactory receptor protein for odorant pyrazines. Proc Natl Acad Sci USA 82:3050–3054, 1985.

21. Pevsner J, Reed RR, Feinstein PG, Snyder SH: Molecular cloning of odorant-binding protein: Member of a ligand carrier family. Science 241:336–338, 1988.

22. Pevsner J, Hou V, Snyder SH: Odorant binding protein: Characterization of ligand binding. J Biol Chem 265:6118–6125, 1990.

23. Ronnett GV, SH Snyder: Molecular messengers of olfaction. Trends Neurosci 15:508–513, 1992.

24. Pevsner J, Sklar PB, Snyder SH: Odorant-binding protein: Localization to nasal glands and secretions. Proc Natl Acad Sci USA 83:4942–4946, 1986.

25. Stroop WG: Viruses and the olfactory system. In: Doty RL (ed): Handbook of Olfaction and Gustation. New York: Marcel Dekker, 1995.

26. Mellert TK, Getchell ML, Sparks L, Getchell TV: Characterization of the immune barrier in human olfactory mucosa. Otolaryngol Head Neck Surg 106:181–188, 1992.

27. Carr WES, Gleeson RA, Trapido-Rosenthal HG: The role of perireceptor events in chemosensory processes. Trends Neurosci 13:212–215, 1990.

28. Dahl AR, Hadley WM: Nasal cavity enzymes involved in xenobiotic metabolism: Effects of the toxicity of inhalants. Crit Rev Toxicol 32:383–407, 1991.

29. Ottoson D: Analysis of the electrical activity of the olfactory epithelium. Acta Physiol Scand 35(suppl 122):1–83, 1956.

30. Gesteland RC, Lettvin JY, Pitts WH: Chemical transmission in the nose of the frog. J Physiol 181:525–559, 1965.

31. Moulton DG: Spatial patterning of response to odors in the peripheral olfactory system. Physiol Rev 56:578–593, 1976.

32. Mackay-Sim A, Shaman P, Moulton DG: Topographic coding of olfactory quality: Odorant-specific patterns of epithelial responsivity in the salamander. J Neurophysiol 48:584–596, 1982.

33. Mozell MM: Evidence for sorption as a mechanism of the olfactory analysis of vapors. Nature 203:1181–1182, 1964.

34. Potter H, Chorover SL: Response plasticity in hamster olfactory bulb: Peripheral and central processes. Brain Res 116:417–429, 1976.

35. Lancet D, Kauer JS, Greer CA, Shepherd GM: High resolution 2-deoxyglucose localization in olfactory epithelium. Chem Sens 6:343–349, 1981.

36. Land LJ: Localized projection of olfactory nerves to rabbit olfactory bulb. Brain Res 63:153–166, 1973.

37. Land LJ, Shepherd GM: Autoradiographic analysis of olfactory receptor projections in the rabbit. Brain Res 70:506–510, 1974.

38. Le Gros Clark WE: The projection of the olfactory epithelium on the olfactory bulb in the rabbit. J Neurol Neurosurg Psychiaty 14:1–10, 1951.

39. Le Gros Clark WE: Inquiries into the anatomical basis of olfactory discrimination. Proc R Soc London 146(B):299–319, 1956.

40. Costanzo RM, Mozell MM: Electrophysiological evidence for a topographical projection of the nasal mucosa onto the olfactory bulb of the frog. J Gen Physiol 68:297–312, 1976.

41. Astic L, Saucier D: Anatomical mapping of the neuroepithelial projection to the olfactory bulb in the rat. Brain Res Bull 16:445–454, 1986.

42. Duncan HJ, Nickell WT, Shipley MT, Gesteland RC: Organization of projections from olfactory epithelium to olfactory bulb in the frog, Rana pipiens. J Comp Neurol 299:299–311, 1990.

43. Gesteland RC, Lettvin JY, Pitts WH, Rojas A: Odor specificities of the frog's olfactory receptors. In: Zotterman Y (ed): Olfaction and Taste, vol. I. London: Pergamon, 1963.

44. Getchell TV: Analysis of unitary spikes recorded extracellularly from frog olfactory receptor cells and axons. J Physiol 234:533, 1973.

45. Getchell TV: Functional properties of vertebrate olfactory receptor neurons. Physiol Rev 66:772–818, 1986.

46. O'Connell RJ, Mozell MM: Quantitative stimulation of frog olfactory receptors. J Neurophysiol 32:51–63, 1969.

47. van Drongelen W: Unitary recordings of near threshold responses of receptor cells in the olfactory mucosa of the frog. J Physiol 277:423–435, 1978.

48. Sicard G, Holley A: Receptor cell responses to odorants: Similarities and differences among odorants. Brain Res 292:283–296, 1984.

49. Holley AA, Duchamp MF, Revial A, Juge A, MacLeod P: Qualitative and quantitative discrimination in the frog olfactory receptors: Analysis from electrophysiological data. Ann NY Acad Sci 237:102–114, 1974.

50. Schild D: Response pattern features of mitral cells in the goldfish olfactory bulb. Brain Res 405:364–370, 1987.
51. Getchell TV, Shepherd GM: Responses of olfactory receptor cells to step pulses of odour at different concentrations in the salamander. J Physiol 282:521–540, 1978.
52. Masukawa LM, Kauer JS, Shepherd GM: Intracellular recordings from two cell types in an in vitro preparation of the salamander olfactory epithelium. Neurosci Lett 35:59–64, 1983.
53. Trotier D, MacLeod P: Intracellular recordings from salamander olfactory receptor cells. Brain Res 268:225–237, 1983.
54. Neher E: The use of the patch clamp technique to study second messenger-mediated cellular events. Neuroscience 26:727–734, 1988.
55. Neher E: Ion channels for communication between and within cells. Science 256:498–502, 1992.
56. Sakmann S: Elementary steps in synaptic transmission revealed by currents through single ion channels. Science 256:503–512, 1992.
57. Firestein S, Werblin F: Gated currents in isolated olfactory receptor neurons of the larval tiger salamander. Proc Natl Acad Sci USA 84:6292–6296, 1987.
58. Schild D: Whole-cell currents in olfactory receptor cells of Xenopus laevis. Exp Brain Res 78:223–232, 1989.
59. Trotier D: A patch-clamp analysis of membrane currents in salamander olfactory receptor cells. Pflugers Arch 407:589–595, 1986.
60. Firestein S, Werblin F: Odor-induced membrane currents in vertebrate-olfactory receptor neurons. Science 244:79–82, 1989.
61. Firestein S, Shepherd GM, Werblin FS: Time course of the membrane current underlying the sensory transduction in salamander olfactory receptor neurones. J Physiol 430:135–158, 1990.
62. Nomura T, Kurihara K: Liposomes as a model for olfactory cells: Changes in membrane potential in response to various odorants. Biochemistry 26:6135–6140, 1987.
63. Wysocki CJ, Beauchamp GK: Ability to smell androstenone is genetically determined. Proc Natl Acad Sci USA 81:4899–4902, 1984.
64. Whissell-Buechy D, Amoore JE: Odour-blindness to musk: Simple recessive inheritance. Nature 242:271–273, 1973.
65. Polak E, Fombon AM, Tilquin C, Punter PH: Sensory evidence for olfactory receptors with opposite chiral selectivity. Behav Brain Res 31:199–206, 1989.
66. Cagan RH, Zeiger WN: Biochemical studies of olfaction: Binding specificity of radioactively labeled stimuli to an isolated olfactory preparation from rainbow trout (Salmo gairdueri). Proc Natl Acad Sci USA 75:4679–4683, 1978.
67. Ludwig J, Margalit T, Eismann E, Lancet D, Kaupp UB: Primary structure of cAMP-gated channel from bovine olfactory epithelium. FEBS Lett 270:24–29, 1990.
68. Reed RR: Signaling pathways in odorant detection. Neuron 8:205–209, 1992.
69. Sklar PB, Anholt RR, Snyder SH: The odorant-sensitive adenylate cyclase of olfactory receptor cells. Differential stimulation by distinct classes of odorants. J Biol Chem 261:15538–15543, 1986.
70. Pace U, Hanski E, Salomon Y, Lancet D: Odorant-sensitive adenylate cyclase may mediate olfactory reception. Nature 316:255–258, 1985.
71. Firestein S, Zufall F, Shepherd GM: Single odor sensitive channels in olfactory receptor neurons are also gated by cyclic nucleotides. J Neurosci 11:3565–3572, 1991.
72. Firestein S, Zufall F, Shepherd GM: Activation of the sensory current in salamander olfactory receptor neurons depends on a G-protein mediated cAMP second messenger system. Neuron 6:825–835, 1991.
73. Kleene SJ, Gesteland RC, Bryant SH: An electrophysiological survey of frog olfactory cilia. J Exp Biol 195:307–328, 1994.
74. Nakamura T, Gold GH: A cyclic nucleotide-gated conductance in olfactory receptor cilia. Nature 325:442–444, 1987.
75. Zufall F, Firestein S, Shepherd GM: Analysis of single cyclic nucleotide gated channels in olfactory receptor cells. J Neurosci 11:3573–3580, 1991.
76. Zufall F, Firestein S, Shepherd GM: Inhibition of the olfactory cyclic nucleotide gated ion channel by intracellular calcium. Proc R Soc London 246(B):225–230, 1991.
77. Fadool DA, Ache BW: Plasma membrane inositol 1,4,5-triphosphate-activated channels mediate signal transduction in lobster olfactory receptor neurons. Neuron 9:907–918, 1992.
78. Hatt H, Ache BW: Cyclic nucleotide- and inositol phosphate-gated ion channels in lobster olfactory receptor neurons. Proc Natl Acad Sic USA 91:6264–6268, 1994.
79. Ache BW: Towards a common strategy for transducing olfactory information. Semin Cell Biol 5:55–63, 1994.
80. Buck L, Axel R: A novel multigene family may encode odorant receptors: A molecular basis for odor recognition. Cell 65:175–187, 1991.
81. Axel R: The molecular logic of smell. Sci Am 273:154–159, 1995.

82. Shepherd GM: Discrimination of molecular signals by the olfactory receptor neuron. Neuron 13:771–790, 1994.
83. Mori K, Yoshihara Y: Molecular recognition and olfactory processing in the mammalian olfactory system. Prog Neurobiol 45:585–619, 1995.
84. Firestein S, Darrow B, Shepherd GM: Activation of the sensory current in salamander olfactory receptor neurons depends on a G protein-mediated cAMP messenger system. Neuron 6:825–835, 1991.
85. Lowe G, Gold GH: Contribution of the ciliary cyclic nucleotide-gated conductance to olfactory transduction in the salamander. J Physiol 462:175–196, 1993.
86. Schiffman SS: Contributions to the physiochemical dimensions of odor: A psychophysical approach. Ann NY Acad Sci 237:164–183, 1974.
87. Schiffman SS: Physicochemical correlates of olfactory quality. Science 185:112–117, 1974.
88. Moulton DG: Spatial patterning of response to odors in the peripheral olfactory system. Physiol Rev 56:578–593, 1976.
89. Mozell MM: Evidence for sorption as a mechanism of the olfactory analysis of vapors. Nature 203:1181–1182, 1964.
90. Mozell MM: The spatio-temporal analysis of odor at the level of the olfactory receptor sheet. J Gen Physiol 50:25–41, 1966.
91. Mozell MM: Evidence for a chromatographic model of olfaction. J Gen Physiol 56:46–63, 1970.
92. Gesteland RC: Neural coding in olfactory receptor cells. In: Beidler LM (ed): Handbook of Sensory Physiology. Chemical Senses, vol. 1, part 1. Berlin: Springer-Verlag, 1971.
93. Amoore JE: Evidence for the chemical olfactory code in man. Ann NY Acad Sci 237:137–143, 1974.
94. Amoore JE, Johnston JW Jr, Rubin M: The stereochemical theory of odor. Sci Am 210:42–49, 1974.
95. Amoore JE: Specific anosmias. In: Getchell TV, Doty RL, Bartoshuk LM, Snow JB (eds): Smell and Taste in Health and Disease. New York: Raven Press, 1991.
96. Ngai J, Chess A, Dowling MM, Necles N, Macagno ER, Axel R: Coding of olfactory information: Topography of odorant receptor expression in the catfish olfactory epithelium. Cell 72:667–680, 1993.
97. Ngai J, Dowling MM, Buck L, Axel R, Chess A: The family of genes encoding odorant receptors in channel catfish. Cell 72:657–666, 1993.

4

Practical Approaches to Clinical Olfactory Testing

RICHARD L. DOTY, Ph.D.

Numerous tests have been devised to assess the ability to smell, although many are too unreliable or time consuming to be practical in the clinical setting.[1] A popular clinical means for assessing smell function has been to ask a patient to sniff small vials containing one or two odorants, such as coffee or cinnamon, and to report whether or not an odor is perceived. Unfortunately, this procedure is akin to testing audition by sounding a horn in a patient's ear and asking whether or not it is heard. This problem is not corrected by having the patient attempt to identify the odors, since without cuing even normal subjects can have difficulty identifying some odors.

During the last two decades remarkable progress has been made in the development of reliable, valid, and clinically applicable olfactory tests. Impetus for this development has come from several sources, including growing public awareness of the importance of this sense for general well-being and resultant increases in governmental funding of clinical chemosensory research, particularly in the United States and Japan. In part because of increased litigation, physicians and insurance carriers are now, more than ever, aware that objective chemosensory assessment is essential for: (1) establishing the validity of a patient's complaint, (2) characterizing the specific nature of the chemosensory problem, (3) accurately monitoring medical or surgical interventions, (4) detecting malingering, (5) counseling patients to cope with their problem, and (6) assigning disability compensation. Evidence that olfactory testing may be of assistance in the differential diagnosis of some neurological diseases will likely add to the importance of quantitative olfactory testing in medical practice.[2-6]

In this chapter I provide an overview of practical psychophysical methods for clinically assessing the ability to smell. Although electrophysiological methods are now available for such assessment, they are still largely experimental and appear to be less sensitive than psychophysical methods to many olfactory deficits; thus, they are not discussed in this chapter (for review, see Doty and Kobal[1]). Also addressed in this chapter are basic issues related to olfactory test reliability and validity, relationships among different types of olfactory tests, and factors that contribute to olfactory loss within the nonclinical population.

STIMULUS CONTROL AND PRESENTATION

Contrary to popular opinion, knowledge of the number of molecules that enter the nose is not needed to make an olfactory test valid. The key point is that the stimuli are reliably generated and presented, and that norms are available to establish whether a patient's test score is normal or abnormal. Thus, accurate clinical assessment of chemosensory function can be made using surprisingly simple stimulus presentation equipment. Historically, such equipment has included: (1) the draw tube olfactometer of Zwaardemaker,[7] (2) glass sniff bottles,[8,9] (3) glass rods, wooden sticks, or strips of blotter paper dipped in various concentrations of stimuli,[10,11] (4) plastic squeeze bottles,[12,13] (5) air-dilution olfactometers,[14–16] (6) microencapsulated "scratch and sniff" odorized strips,[17] and (7) bottles from which blasts of odorant-saturated air are presented[18] (Fig. 4–1). Although intravenous odorant administration (eg, injection of thiamine propyldisulfide into the cubital vein) also has been used to produce chemosensory sensations, the assumption that the blood-borne stimulus makes its way to the olfactory receptors via diffusion from capillary bodies is controversial.[19]

PSYCHOPHYSICAL TEST PROCEDURES

Psychophysical tests provide a quantitative measure of sensory function obtained from verbal or other conscious responses of a subject to perceptual events related to the presentation of stimuli. Classical psychophysical procedures, such as those used for establishing threshold values (see below), were developed in the 19th century by German investigators, most notably Heinrich Weber[20] and Gustof Fechner.[21] Today such procedures are the primary means for assessing sensory function in not only olfactometry, but also audiometry and optometry (for detailed treatises on psychophysical methods, see references 22 to 26).

The most widely used psychophysical tests of olfactory function are those in which: (1) the sensitivity of the sensory system to low concentrations of odorants is measured (eg, detection threshold tests), and (2) the ability to identify odorants is determined when cued responses are made available (eg, forced-choice odor identification tests). The present chapter focuses on these two classes of tests, since such tests: (1) are relatively easy to administer, (2) have been demonstrated to be sensitive to olfactory deficits that accompany a wide variety of disease states, and (3) are of practical use in the clinic. Other types of olfactory tests (eg, tests of odor memory, differential thresholds, odor discrimination, intensity scaling) have yet to prove their usefulness in clinical settings and, for this reason, are not described here. The interested reader is referred elsewhere for more details of such tests.[1,23–25,27,28]

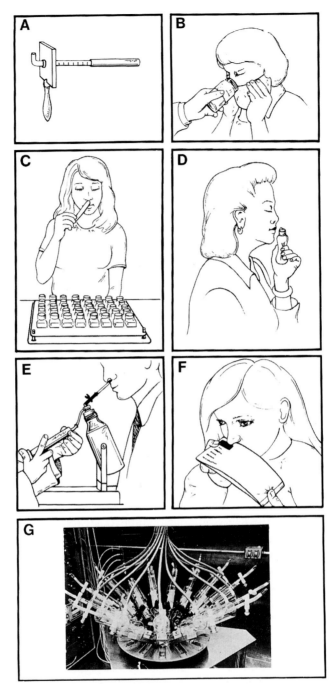

Figure 4–1. Procedures for presenting odorants to subjects for assessment. **A**: An early drawtube olfactometer of Zwaardemaker. In this apparatus, an outer tube made of rubber or another odorous material slides along a calibrated inner tube, one end of which is inserted into the subject's nostril. When the odorized tube is slid towards the subject, less of its internal surface is exposed to the inspired airstream, resulting in a weaker olfactory sensation. **B**: Sniff bottle. **C**: Perfumist's strip. **D**: Squeeze bottle. **E**: Blast injection device. The experimenter injects a given volume of odor into the bottle and releases the pressure by sqeezing a clamp on the tube leading to the nostril, producing a stimulus pulse. **F**: Microencapsulated "scratch and sniff" University of Pennsylvania Smell Identification Test. **G**: Sniff ports on a rotating table connected to the University of Pennsylvania's Dynamic Air-Dilution Olfactometer. From Doty and Kobal.[1]

Threshold Tests

A popular means for assessing olfactory function is to establish, operationally, a measure of the lowest concentration of a stimulus that can be discerned, ie, the absolute threshold. A qualitative odor sensation is rarely perceived at such low concentrations (eg, "roselike"), where only the faint presence of something vs nothing is noticed. The *detection threshold* is the lowest odorant concentration where such a presence is reliably detected, whereas the *recognition threshold* is the lowest concentration where odor quality is reliably discerned.[1] Since the later thresholds are often confounded with response biases,[29,30] they are rarely used clinically and, therefore, they are not further described in this chapter.

In the case of detection threshold testing, a patient may be asked to indicate, on a given trial, which of two or more stimuli (eg, an odorant and one or more blanks) smells strongest, rather than to report whether an odor is perceived or not. Such "forced-choice" procedures are believed to be less susceptible than non–forced choice procedures to contamination by response criteria (ie, the liberalism or conservatism in reporting the presence or absence of an odor under uncertain conditions) and typically produce lower threshold values.[31] Although other measures, such as ones derived from signal detection theory, can theoretically separate sensory sensitivity from response bias, they require a large number of trials and therefore have less clinical utility.

Two classes of detection threshold procedures commonly used in the clinic are the *ascending method of limits* (AML) and the *single staircase* (SS). In the AML procedure, stimuli are increased in concentration from a level not initially smelled by the subject to concentration levels where detection is reliably made. Typically, a point of transition between no detection and detection is determined. With the exception of the Japanese T&T olfactometer,[32] modern olfactory tests require forced-choice responses on each trial. In the SS method (a variant of the method of limits technique; see Cornsweet[33]), the stimulus concentration is increased following trials on which a subject fails to detect the stimulus and decreased following trials where correct detection occurs. In both the AML and SS procedures, the stimulus presentations are made from weak to strong in an effort to reduce potential adaptation effects of prior stimulation.[34]

At our center we use a SS procedure that incorporates phenyl ethyl alcohol (PEA), a rose-smelling odorant that has minimal propensity to stimulate intranasal free nerve endings from the trigeminal nerve.[35] A given trial consists of the presentation of two 100-mL glass sniff bottles to the patient in rapid succession (see Fig. 4–1A). One bottle contains 20 mL of a given concentration of PEA dissolved in light mineral oil, whereas the other contains only mineral oil. The patient is asked to report which of the two bottles in a pair produces the strongest sensation. The first trial is presented at a −6.50 log (liquid volume/volume) concentration step. If a miss occurs on any trial before five are correctly completed at a given concentration, the process is repeated at one log concentration step higher. When five consecutive correct trials occur at a given concentration level, the staircase is "reversed" and the next pair of trials is presented 0.5 log concentration step lower. From this point on, only one or two trials are presented at each step (ie, if the first trial is missed, the second is not given and the staircase is moved to the next higher 0.5 log step concentration). When correct performance occurs on both trials, the next trial is given 0.5 concentration step lower. The geometric mean of the last four of seven staircase reversal points is the threshold estimate. Examples of individual threshold data collected in this manner are shown in Figure 4–2.

Quality Identification Tests

Among the most popular procedures for assessing taste and smell function are those that require stimulus quality identification, and include *odor-naming tests, yes/no identification tests*, and *multiple-choice identification tests*. The responses required, on a given trial, for each of these respective classes of tests are: (1) to provide a name for the stimulus, (2) to indicate whether or not the stimulus smells like an object

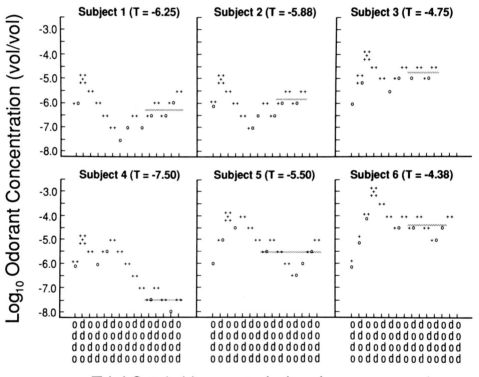

Trial Set (with counterbalancing sequence)

Figure 4–2.   Data illustrating single-staircase detection threshold determinations. Each plus (+) indicates a correct detection when an odorant vs a blank is presented. Each minus (−) indicates incorrect report of an odorant. Threshold value (T; vol/vol in light USP grade mineral oil) is calculated as the mean of the last four of seven staircase reversals. The *o*s and *d*s on the abcissa indicate the counterbalancing order of the presentation sequences for each trial and are read downward (o = odorant presented first, then diluent; d = diluent presented first, then odorant). At the first reversal point (where five correct sets of pairs occur at the same concentration), the fifth order sequence is determined by the first *o* or *d* of the subsequent column of four order sequences. Although these data represent staircases begun at the −6.00 log vol/vol concentration, we now typically begin the procedure at the −6.50 log concentration. From Doty.[28]

named by the examiner (eg, does this smell like a rose?), and (3) to identify the stimulus from a list of names.

Odor-naming tests require that the patient provide a name for the odorant that is presented. Although used clinically,[36] such tests suffer from the problem that many normal individuals have difficulty in naming or identifying even familiar odors without cuing. In yes/no identification tests, this problem is circumvented, in that a patient must report whether each of a set of stimuli smells like a particular substance named by the experimenter. On a given trial, two stimuli are commonly presented, with the correct alternative provided on one trial and an incorrect one on the other (eg, rose odor is presented and the subject is asked on one trial whether the odor smells like rose and on another trial whether the odor smells like peppermint).[37]

A number of multiple-choice odor identification tests have been described in the clinical literature.[17,23,28,38–40] These tests are conceptually similar and, in the few cases that have been examined, strongly correlated with one another.[40–42] The most widely used of these tests (the University of Pennsylvania Smell Identification Test [UPSIT], commercially termed the Smell Identification Test™, Sensonics, Inc., Haddon Heights, NJ)[17,43,44] was developed in our laboratory and evaluates the ability of patients to identify, from sets of four descriptors, each of 40 "scratch and sniff" odorants (see Fig. 4–1F). The number of items correctly identified out of 40 serves as the test measure. This value is compared to norms based upon data from a large number of individuals sampled from the community at large and a percentile rank is determined, depending upon the age and gender of the patient.[43] This test is amenable to self-administration and provides a means for detecting malingering. Over a hundred publications have appeared using this test, which has been found to be sensitive to olfactory dysfunction associated with many medical disorders, including Alzheimer's disease, amyotrophic lateral sclerosis, Huntington's disease, Kallmann's syndrome, Korsakoff's psychosis, Parkinson's disease, extreme low-weight anorexia, nasal sinus disease, and schizophrenia.[2,45–50]

Several odor identification tests have incorporated "confusion matrices."[23,40] In the most popular of these tests,[40] each of 10 suprathreshold stimuli are presented to a patient in counterbalanced order 10 times (100 total trials). The response alternatives are the names of the 10 stimuli: ammonia, Clorox, licorice, mothballs, peppermint, roses, turpentine, vanilla, Vicks vapor rub, and vinegar. No feedback as to the correctness of the subjects' responses is given. The percentage of responses given to each alternative for each odorant are determined and displayed in a rectangular matrix (stimuli making up rows and response alternatives making up equivalently ordered columns). Responses along the negative diagonal therefore represent correct responses, whereas those which fall away from the diagonal represent "confusions." The percentage of correct responses is used as the main test measure, although research is currently being performed to see if the confusions (off-diagonal responses) may provide clinically meaningful information.

ISSUES RELATED TO TEST RELIABILITY AND VALIDITY

The usefulness of a test of sensory function depends upon its reliability (eg, its consistency of measurement) and validity (the degree to which it measures what it portends to measure). Related to a test's validity are its sensitivity (ability to detect abnormalities when present) and specificity (ability to detect such abnormalities with a minimum of false positives).[49]

Several techniques for estimating reliability have been developed that are applicable to olfactory tests. For example, a test can be administered on two occasions to a group of subjects and a correlation coefficient computed between the test scores of the two occasions (termed the *test-retest reliability coefficient* or the coefficient of stability). On the other hand, subsections of a test (eg, odd vs even items of a multiple-item odor identification test) can be correlated with one another (termed the *split-half reliability coefficient*). Since reliability is related to test length, a statistical correction for test length must be applied to the correlation coefficient obtained in this way to obtain the correct reliability coefficient for the full test.[22]

A test's validity is usually more difficult to determine mathematically than its reliability. Forms of validity include *content validity* (the degree to which a test is made up of items, odorants, tastants, etc relevant to the task at hand) and *criterion-related validity* (the degree to which a test correlates with criteria known to be associated with the trait to be measured).

At the present time, information on the reliability or validity of most clinical olfactory tests is limited. In general, forced-choice odor identification tests evidence a high degree of reliability (eg, both the test-retest and split-half *r*s of the UPSIT are above 0.90[51]). Although olfactory detection threshold values vary considerably among individuals and evidence considerable day-to-day fluctuations within the same individuals,[52] they also evidence, with noticeable exceptions,[53] reasonable reliability. Thus, in a study of 40 subjects, Koelega[54] reported test-retest reliability coefficients (Spearman *r*) for a four-alternative forced-choice *n*-amyl acetate threshold test to be 0.65, 0.51, and 0.59 for bilateral, right nostril, and left nostril presentations, respectively. More recently, in a study of 32 subjects ranging in age from 22 to 59 years, Cain and Gent[55] established correlations between the left and right sides of the nose for detection threshold values of butanol, phenyl ethyl methyl carbinol, isoamyl butyrate, and pyridine. The high correlations (0.68, 0.96, 0.86, and 0.83, respectively) were interpreted as reflecting split-half reliability coefficients. Recently, we evaluated the test-retest reliability of 10 olfactory tests in the same set of 57 subjects aged 24 to 84 years.[42] Most of the olfactory tests, including the threshold tests, evidenced relatively high test-retest reliability, with the median reliability coefficient being 0.70.

WHAT DO OLFACTORY TESTS REALLY MEASURE?

Implicit in the application of any olfactory test is the assumption that it measures a specific sensory attribute associated with its name. For example, an "odor memory test" is assumed to measure the ability to remember odors and to activate cortical circuits involved in odor memory processing. Likewise, a test of odor identification is presumed to measure a reasonably specific attribute that relates to the ability to identify odors and to tap neural circuits subserving this attribute.

Despite the apparent logic underlying this assumption, little is known about the degree to which different olfactory tests measure unique perceptual attributes or neural substrates. Importantly, in some situations the logic of this assumption breaks down and misleading inferences can result from literal interpretations of test results. For example, if one has little ability to detect or recognize odorants, then poor performance would be observed on an odor memory test, even though underlying neural circuits necessary for odor memory per se may be intact. In this case, the score on the odor memory test is not a valid reflection of the substrate of odor memory per se, since no olfactory sensation to be remembered is available for encoding.

Recently we performed a principal components analysis of the correlations among a set of 13 measures derived from 9 diverse olfactory tests (eg, nominal tests of odor identification, discrimination, identification, etc) to determine the degree to which separate principal components emerged.[42] The subjects were comprised of persons from the general population whose ages varied widely. The analysis revealed four meaningful components, the first consisting of strong primary loadings from 9 of the 13 measures, including measures of odor threshold, identification, and memory. The second component received strong primary loadings derived from a suprathreshold test of odor intensity, whereas the third was comprised of a primary loading of the average pleasantness values given to this same set of suprathreshold stimuli. The fourth component consisted mainly of a loading from the bias measure of a yes/no odor identification task. These findings suggest that most of the tests evaluated measure, to a large degree, a common source of variance.[42,56]

Such findings bring to mind the days when psychological tests were initially classified under categories assumed to assess separate mental functions.[57] However, this assumption was later found to be questionable. As noted by Guilford[22] (p 471), "Most tests of mental ability exhibit some degree of positive correlation; often two tests that are classified under the same name exhibit no more correlation than do two other tests supposedly belonging to two different categories of ability. The notion of broad unitary powers which operate singly and in an isolated manner must therefore be discarded."[22] The same would appear to be true with many olfactory tests!

OTHER CONSIDERATIONS

Unilateral vs Bilateral Testing

In general, most individuals with chemosensory dysfunction evidence the dysfunction bilaterally, although unilateral deficits are reliably present in some patients, particularly those with asymmetrical brain injury. Furukawa, Kamide, Miwa, and Umeda[58] reported that 7 of 94 patients (7.4%) they examined, all of whom evidenced no bilateral threshold deficits, evidenced significant unilateral threshold deficits. They reported a similar phenomenon in 6 of 12 patients who had undergone brain surgery. Of 82 consecutive nonanosmic patients presenting to the University of Pennsylvania Smell and Taste Center with chemosensory dysfunction, 14 (ie, 17%) were observed whose unilateral detection thresholds were discrepant from one another by at least three orders of magnitude (Doty, unpublished). Interestingly, 9 of these 14 individuals were anosmic on one side of the nose, even though only 3 had bilateral detection threshold values that were obviously abnormal.

Unilateral testing is not difficult. For such testing, it is important to close the contralateral naris without distorting the septum (eg, by using a piece of Microfoam tape [3M Corporation, Minneapolis] cut to fit tightly over the borders of the naris) and have the patient exhale through the mouth after inhaling through the nose.[59] This precaution serves, as in the case when both nares are blocked, to insure that odor-laden air does not enter the contralateral nasal chamber via the retronasal route during testing.

Detection of Malingering

Because compensation is often available from insurance companies for olfactory dysfunction caused by accidents, malingering on chemosensory tasks sometimes occurs. It is commonly assumed that if a patient cannot readily perceive the vapors from an irritating substance presented to the nose, he or she is malingering.[60] While this may be true in some cases, this is not a definitive test of malingering, since: (1) individuals who, on other grounds, are believed to be feigning anosmia have difficulty in denying experiencing the effects of NH_4 or other irritants, particularly since these stimuli often produce overt reflexes such as eye watering and coughing, (2) the trigeminal nerve can undergo damage in some types of head injury, and (3) considerable variability exists among normal individuals in trigeminal responsiveness to such irritants.

Another approach for detecting malingering relates to response strategies employed by patients on forced-choice tests, since malingerers often avoid the correct response more often than expected on the basis of chance. This is well illustrated by responses to the UPSIT. Since the UPSIT is a four-alternative forced-choice test, approximately 25% of the test items (ie, 10) are correctly answered, on average, by an anosmic. The probability of scoring 5 or less on the UPSIT and not having at least some ability to smell is less than 5 in 100. The probability of scoring 0 on the UPSIT and having no sense of smell is 1 in 10 000.[43] Simulation studies demonstrate that about two thirds of college students tested, when asked to fool an examiner that they have no sense of smell, avoid the correct answer at a level beyond that expected on the basis of random responding.[17]

Major Subject Variables That Influence Olfactory Ability

As noted in other chapters of this book, numerous subject factors are related to the ability to smell, including—in order of relative importance—age, gender, and smoking.[28,61,62] Markedly decreased olfactory function is observed in nearly three quarters of the population over the age of 80, and in approximately one half of the population between the ages of 65 and 80.[44] In addition, age influences the responsiveness of the nasal mucosa to volatile chemicals that produce irritation and other skin sensations via trigeminal free nerve endings.[63] Additionally, large individual differences are present in the test scores of older persons and, on average, men evidence age-related declines in odor perception at an earlier age than do women. Importantly, such declines are not inconsequential and may be the basis, in part, for why a disproportionate number of elderly persons die from accidental gas poisoning[64] and why many complain that food lacks flavor.[44] The latter phenomenon, which can lead to complacency in eating, may be related to some cases of age-related nutritional deficiencies. As observed clinically,[48] decreased "taste" perception during deglutition usually reflects the loss of retronasal stimulation of the olfactory receptors.[65,66]

In general, women have a better sense of smell than men, as reflected by test scores on a variety of olfactory tests.[17,44,67–73] Such sex differences are evident soon after birth and may be due to the early influences of gonadal hormones on brain organization and the general development of sexually dimorphic traits.[74]

Tobacco smoking influences the ability to smell although, surprisingly, such influences, on average, are not as marked as those of either age or gender.[44] These alterations are dose related and present in both current and past smokers.[75] Cessation from smoking results in some improvement of olfactory function—improvement that is related to the dose of previous smoking and the duration of such cessation.

As with the case of smoking, changes in smell function have been observed following exposure to a wide variety of airborne environmental agents, including industrial chemicals and dusts.[76] In the largest epidemiological study to date on this topic, the UPSIT was administered to 731 workers at a chemical plant that manufactures acrylates and methacrylates.[77] Decreased olfactory function proportionate to the estimated dose exposure levels to these acrylates was found. Interestingly, a seemingly paradoxical interaction between smoking habits and acrylate-related olfactory dysfunction was found. Thus, individuals who had never smoked cigarettes but who had been exposed to the acrylates were six times more likely than their smoking counterparts to evidence olfactory decrements. This suggests the possibility that smoking behavior protected the olfactory system from damage from the acylates, perhaps by increasing protective enzymatic activity within the olfactory neuroepithelium.

Prior experience with odors influences a number of olfactory test measures, including measures of sensitivity. For example, repeated testing within the perithreshold concentration range results in decreased thresholds or enhancement of signal detection sensitivity measures.[78–81] Training improves performance on a wide variety of olfactory tests, including tests of odor identification, detection, and discrimination.[82–84] Interestingly, the hedonic quality of odorants can be influenced by repeated exposure, making unpleasant odors less unpleasant and pleasant odors less pleasant.[85] Assuming that adaptation or habituation is not the primary basis for this phenomenon, affective components of odors apparently can be altered independently of odor intensity.

CONCLUSIONS

Both theoretical and practical aspects of psychophysical olfactory testing have been presented in this chapter. The fact that a number of types of olfactory tests appear to measure the same general elements of olfactory functioning, at least in "normal" populations consisting of both young and elderly individuals, suggests that the clinician has wide choice in selecting tests from which to evaluate the ability to smell. Such factors as the ease and cost of administration, the availability of normative data, and degree to which there is a need to detect subtle degrees of dysfunction should be the basis for the selection of olfactory tests for a given application.

As alluded to earlier in the chapter, work is sorely needed to establish the comparative sensitivity of olfactory tests to various disease states because, in some disorders, early detection of subtle dysfunction may be of considerable diagnostic or prognostic value. For example, early detection of Parkinson's disease may result in earlier and thus more effective treatment with monoamine oxidase inhibitors and other agents that may retard disease progression or mitigate symptom severity.[86]

Acknowledgments—Supported by Grant PO1 DC 00161 from the National Institute on Deafness and Other Communication Disorders and Grant RO1 AG 08148 from the National Institute on Aging.

REFERENCES

1. Doty RL, Kobal G: Current trends in the measurement of olfactory function. In: Doty RL (ed): Handbook of Olfaction and Gustation. New York: Marcel Dekker, pp 191–225, 1995.
2. Doty RL: Olfactory dysfunction in neurodegenerative disorders. In: Getchell TV, Doty RL, Bartoshuk LM, Snow JB (eds): The Chemical Senses in Health and Disease. New York: Raven, 1991.
3. Doty RL, Singh A, Tetrude J, Langston JW: Lack of major olfactory dysfunction in MPTP-induced parkinsonism. Ann Neurol 32:97–100, 1992.
4. Sajjadian A, Doty RL, Gutnick DN, Chirurgi RJ, Sivak M, Perl D: Olfactory dysfunction in amyotrophic lateral sclerosis. Neurodegeneration 3:153–157, 1994.
5. Gregson RAM, Smith DAR: The clinical assessment of olfaction: Differential diagnoses including Kallmann's syndrome. J Psychosomat Res 25:165–174, 1981.
6. Stern MB, Doty RL, Dotti M, et al: Olfactory function in Parkinson's disease subtypes. Neurology 44:266–268, 1994.
7. Zwaardemaker H: L'Odorat. Paris: Doin, 1925.
8. Cheesman GH, Townsend MJ: Further experiments on the olfactory thresholds of pure chemical substances, using the "sniff-bottle method." Q J Exp Psychol 8:8–14, 1956.
9. Doty RL, Gregor T, Settle RG: Influences of intertrial interval and sniff bottle volume on the phenyl ethyl alcohol olfactory detection threshold. Chem Senses 11:259–264, 1986.
10. Semb G: The detectability of the odor of butanol. Percept Psychophys 4:335–340, 1968.
11. Toyota B, Kitamura T, Takagi SF: Olfactory Disorders—Olfactometry and Therapy. Tokyo: Igaku-Shoin, 1978.
12. Amoore JE, Ollman BG: Practical test kits for quantitatively evaluating the sense of smell. Rhinology 21:49–54, 1983.
13. Cain WS, Gent JP, Goodspeed RB, Leondard G: Evaluation of olfactory dysfunction in the Connecticut Chemosensory Clinical Research Center. Laryngoscope 98:83–88, 1988.
14. Doty RL, Deems DA, Frye R, Pelberg R, Shapiro A: Olfactory sensitivity, nasal resistance, and autonomic function in the multiple chemical sensitivities (MCS) syndrome. Arch Otolaryngol Head Neck Surg 114:1422–1427, 1988.
15. Punter PH: Measurement of human olfactory thresholds for several groups of structurally related compounds. Chem Senses 7:215–235, 1983.
16. Walker JC, Kurtz DB, Shore FM, Ogden MW, Reynolds JH IV: Apparatus for the automated measurement of the responses of humans to odorants. Chem Senses 15:165–177, 1990.
17. Doty RL, Shaman P, Dann M: Development of the University of Pennsylvania Smell Identification Test: A standardized microencapsulated test of olfactory function. Physiol Behav 32:489–502, 1984.

18. Elsberg CA, Levy I: The sense of smell: I. A new and simple method of quantitative olfactometry. Bull Neurol Inst NY 4:5–19, 1935.
19. Maruniak JA, Silver WL, Moulton DG: Olfactory receptors respond to blood-borne odorants. Brain Res 265:312–316, 1983.
20. Weber EH: De pulsu, resorptione, auditu ettactu: Annotationes anatomicae et Physiologicae. Leipsig, Germany: Koehler, 1834.
21. Fechner GT: Elemente der Psychophysik. Leipzig, Germany: Breitkopf and Harterl, 1860.
22. Guilford JP: Psychometric Methods. New York: McGraw-Hill, 1954.
23. Köster EP: Human psychophysics in olfaction. In: Moulton DG, Turk A, Johnston JW Jr (eds): Methods in Olfactory Research. New York: Academic Press, 1975.
24. Marks LE: Sensory Processes. New York: Academic Press, 1974.
25. Stevens SS: The psychophysics of sensory function. In: Rosenblith WA (ed): Sensory Communication. Cambridge, MA: MIT Press, 1961.
26. Tanner WP Jr, Swets JA: A decision-making theory of visual detection. Psychol Rev 61:401–409, 1954.
27. Cain WS, Cometto-Muñiz JE, Wijk RA: Techniques in the quantitative study of human olfaction. In: Serby MJ, Chobor KL (eds): Science of Olfaction. New York: Springer-Verlag, 1993.
28. Doty, RL: Olfactory system. In: Getchell TV, Doty RL, Bartoshuk LM, Snow JB Jr (eds): Smell and Taste in Health and Disease. New York: Raven Press, 1991.
29. O'Mahony M: Gustatory responses to nongustatory stimuli. Perception 12:627–633, 1983.
30. Weiffenbach JM: Taste-quality recognition and forced-choice response. Percept Psychophys 33:251–254, 1983.
31. Blackwell HR: Psychophysical thresholds: Experimental studies of methods of measurement. Bull Engineer Res Inst 36, 1953.
32. Takagi SF: Human Olfaction. Tokyo: Tokyo Press, 1989.
33. Cornsweet TN: The staircase-method in psychophysics. Am J Psychol 75:485–491, 1962.
34. Pangborn RM, Berg HW, Roessler EB, Webb AD: Influence of methodology on olfactory response. Percept Motor Skills 18:91–103, 1964.
35. Doty RL, Brugger WE, Jurs PC, Orndorff MA, Snyder PJ, Lowry LD: Intranasal trigeminal stimulation from odorous volatiles: Psychometric responses from anosmics and normal humans. Physiol Behav 20:175–185, 1978.
36. Gregson RAM, Free ML, Abbott MW: Olfaction in Korsakoffs, alcoholics and normals. Br J Clin Psychol 20:3–10, 1981.
37. Corwin J: Olfactory identification in hemodialysis: Acute and chronic effects on discrimination and response bias. Neuropsychologia 27:513–522, 1989.
38. Cain WS, Gent J, Catalanotto FA, Goodspeed RB: Clinical evaluation of olfaction. Am J Otolaryngol 4:252–256, 1983.
39. Wood JB, Harkins SW: Effects of age, stimulus selection, and retrieval environment on odor identification. J Gerontol 42:584–588, 1987.
40. Wright HN: Characterization of olfactory dysfunction. Arch Otolaryngol Head Neck Surg 113:163–168, 1987.
41. Cain WS, Rabin RD: Comparability of two tests of olfactory functioning. Chem Senses 14:479–485, 1989.
42. Doty RL, Smith R, McKeown D, Raj J: Tests of human olfactory function: Principal components analysis suggests that most measure a common source of variance. Percept Psychophys 56:701–707, 1994.
43. Doty RL: The Smell Identification Test™ Administration Manual, 3rd ed. Haddon Heights, NJ: Sensonics, 1995.
44. Doty RL, Shaman P, Applebaum SL, Giberson R, Sikorsky L, Rosenberg L: Smell identification ability: Changes with age. Science 226:1441–1443, 1984.
45. Doty RL, Golbe LI, McKeown DA, Stern MB, Lehrach CM, Crawford D: Olfactory testing differentiates between progressive supranuclear palsy and idiopathic Parkinson's disease. Neurology 43:962–965, 1993.
46. Doty RL, Deems D, Steller S: Olfactory dysfunction in Parkinson's disease: A general deficit unrelated to neurologic signs, disease stage, or disease duration. Neurology 38:1237–1244, 1988.
47. Doty RL, Reyes P, Gregor T: Presence of both odor identification and detection deficits in Alzheimer's disease. Brain Res Bull 18:597–600, 1987.
48. Deems DA, Doty RL, Settle RG, et al: Smell and taste disorders, a study of 750 patients from the University of Pennsylvania Smell and Taste Center. Arch Otolaryngol Head Neck Surg 117:519–528, 1991.
49. Schwartz, BS: The epidemiology of olfactory dysfunction. In: Laing DG, Doty RL, Breipohl B (eds): The Human Sense of Smell. Berlin: Springer-Verlag, 1991.

50. Fedoroff IC, Stoner SA, Andersen AE, Doty RL, Rolls BJ: Olfactory dysfunction in anorexia and bulimia nervosa. Int J Eat Disorders 18:71–77, 1995.

51. Doty RL, Agrawal U, Frye R: Evaluation of the internal consistency reliability of the fractionated and whole University of Pennsylvania Smell Identification Test. Percept Psychophys 45:381–384, 1989.

52. Stevens JC, Cain WS, Burke RJ: Variability of olfactory thresholds. Chem Senses 13:643–653, 1988.

53. Heywood PG, Costanzo RM: Identifying normosmics: A comparision of two populations. Am J Otolaryngol 7:194–199, 1986.

54. Koelega HS: Olfaction and sensory asymmetry. Chem Senses Flav 4:89–95, 1979.

55. Cain WS, Gent JF: Olfactory sensitivity: Reliability, generality, and association with aging. J Exp Psychol Hum Percept Perf 17:382–391, 1991.

56. Yoshida M: Correlation analysis of detection threshold data for "standard test" odors. Bull Faculty Sci Eng Chuo Univ 27:343–353, 1984.

57. Whipple GM: Manual of Mental and Physical Tests. Baltimore: Warwick and York, 1914.

58. Furukawa M, Kamide M, Miwa T, Umeda R: Importance of unilateral examination in olfactometry. Auris Nasus Larynx (Tokyo) 15:113–116, 1988.

59. Doty RL, Stern MB, Pfeiffer C, Gollomp SM, Hurtig HI: Bilateral olfactory dysfunction in early stage treated and untreated idiopathic Parkinson's disease. J Neurol Neurosurg Psychiat 55:138–142, 1992.

60. Griffith IP: Abnormalities of smell and taste. Practitioner 217:907–913, 1976.

61. Doty RL, Snow JB Jr: Age-related changes in olfactory function. In: Margolis RL, Getchell TV (eds): Molecular Neurobiology of the Olfactory System. New York: Plenum Press, 1988.

62. Schiffman SS: Olfaction in aging and medical disorders. In: Serby MJ, Chobor KL (eds): Science of Olfaction. New York: Springer-Verlag, 1993.

63. Stevens JC, Cain WS: Smelling via the mouth: Effect of aging. Percept Psychophys 40:142–146, 1986.

64. Chalke HD, Dewhurst JR, Ward CW: Loss of sense of smell in old people. Public Health 72:223–230, 1958.

65. Burdach KJ, Doty RL: The effects of mouth movements, swallowing, and spitting on retronasal odor perception. Physiol Behav 41:353–356, 1987.

66. Mozell MM, Smith BP, Smith PE, Sullivan RL Jr, Swender P: Nasal chemoreception in flavor identification. Arch Otolaryngol 90:131–137, 1969.

67. Cain WS: Odor identification by males and females: Predictions vs. performance. Chem Senses 7:129–142, 1982.

68. Doty RL, Ford M, Preti G, Huggins G: Human vaginal odors change in pleasantness and intensity during the menstrual cycle. Science 190:1316–1318, 1975.

69. Doty RL, Kligman A, Leyden J, Orndorff MM: Communication of gender from human axillary odors: Relationship to perceived intensity and hedonicity. Behav Biol 23:373–380, 1978.

70. Doty RL, Ram CA, Green P, Yankell S: Communication of gender from breath odors: Relationship to perceived intensity and pleasantness. Horm Behav 16:13–22, 1982.

71. Doty RL: Gender and endocrine-related influences upon olfactory sensitivity. In: Meiselman HL, Rivlin RS (eds): Clinical Measurement of Taste and Smell. New York: MacMillan, 1986.

72. Koelega HS, Köster EP: Some experiments on sex differences in odor perception. Ann NY Acad Sci 237:234–246, 1974.

73. Le Magnen J: Les Phenomenes olfacto-sexuels chez l'homme. Arch Sci Physiol 6:125–160, 1952.

74. Makin JW, Porter RH: Attractiveness of lactating females' breast odors to neonates. Child Dev 60:803–810, 1989.

75. Frye RE, Schwartz B, Doty RL: Dose-related effects of cigarette smoking on olfactory function. JAMA 263:1233–1236, 1990.

76. Amoore JE: Effects of chemical exposure on olfaction in humans. In: Barrow CS (ed): Toxicology of the Nasal Passages. Washington, DC: Hemisphere Publishing, 1986.

77. Schwartz B, Doty RL, Monroe C, Frye RE, Barker S: The evaluation of olfactory function in chemical workers exposed to acrylic acid and acrylate vapors. Am J Public Health 79:613–618, 1989.

78. Doty RL, Snyder P, Huggins G, Lowry LD: Endocrine, cardiovascular, and psychological correlates of olfactory sensitivity changes during the human menstrual cycle. J Comp Physiol Psychol 95:45–60, 1981.

79. Engen T: Effect of practice and instruction on olfactory thresholds. Percept Motor Skills 10:195–198, 1960.

80. Rabin MD, Cain WS: Determinants of measured olfactory sensitivity. Percept Psychophys 39:281–286, 1986.

81. Wysocki CJ, Dorries KM, Beauchamp GK: Ability to perceive androstenone can be acquired by ostensibly anosmic people. Proc Nat Acad Sci USA 86:7976–7978, 1989.

82. Desor JA, Beauchamp GK: The human capacity to transmit olfactory information. Percept Psychophys 16:551–556, 1974.

83. Smith RS, Doty RL, Burlingame GK, McKeown DA: Smell and taste function in the visually impaired. Percept Psychophys 54:649–655, 1993.
84. Engen T, Ross BM: Long-term memory of odors with and without verbal descriptions. J Exp Psychol 100:221–227, 1973.
85. Cain WS, Johnson F Jr: Lability of odor pleasantness: Influence of mere exposure. Perception 7:459–465, 1978.
86. Langston JW, Koller WC: The next frontier in Parkinson's disease: Pre-symptomatic detection. Neurology 41(suppl 2):5–7, 1991.

5

Olfactory Loss Secondary to Nasal and Sinus Pathology

ALLEN M. SEIDEN, M.D.

Recent surveys indicate that symptoms relating to chronic sinusitis have become more common than those resulting from arthritis and hypertension.[1] Clearly, nasal and sinus diseases account for a large number of patient visits to both an otolaryngology and primary care practice, yet surprisingly little has been written correlating such pathology to olfactory loss. Whereas an audiological assessment routinely precedes any sort of ear surgery, similar function testing for smell prior to nasal surgery does not often occur. In fact, many patients are not even queried regarding their sense of smell.

Nevertheless, nasal and sinus disease is one of the most common causes of olfactory loss, accounting for 15% to 27% of patients presenting to a taste and smell clinic.[2-4] Contrary to a sensory or neural loss as would occur, for example, after a viral infection or head injury respectively, a loss secondary to nasal or sinus pathology is thought to be conductive. In other words, odorant molecules simply cannot reach the olfactory cleft to stimulate the olfactory receptors. Whereas no specific therapies have been found to be effective in the case of a sensorineural loss,[5,6] inflammatory or obstructive abnormalities in the nose impeding olfactory transport should certainly be amenable to further treatment. It is therefore very important that such a diagnosis not be missed.

At first this may not seem to be a diagnostic problem. If a patient presents with obvious nasal obstruction and associated sinusitis, and complains of an associated loss of smell, it is quite logical to assume the loss is conductive. However, such patients will more often present because of an impaired nasal airway, discharge, or headache pain and recognize the loss of smell to be a predictable consequence. On the other hand, in patients presenting specifically because of a conductive olfactory loss, related sinus symptoms may be surprisingly absent.[7] For example, it has been shown that patients may suffer an olfactory loss in the presence of ostiomeatal pathology, yet have no complaint of nasal obstruction.[8] In addition, those patients describing recurrent bouts of sinusitis may at some point have experienced a viral

52

episode that could have precipitated their loss (see chapters 2 and 6), adding confusion as to the appropriate etiology.

Therefore, while it is imperative that symptoms relating to rhinitis and sinusitis be explored, it is also true that some patients will not fit the predicted pattern. Many of these patients have been to a variety of physicians without having been properly diagnosed and without having been afforded proper treatment. This chapter will review some of the more common characteristics of these patients to clarify the nature of a conductive olfactory loss, as well as consider those steps helpful in making the diagnosis.

THE RELATIONSHIP BETWEEN OLFACTORY ACUITY AND NASAL PHYSIOLOGY

To understand how nasal or sinus disease might adversely effect olfactory acuity, it is helpful to explore the functional relationship between nasal airflow and olfaction under normal conditions. The nose, of course, serves several functions, most of which are carried out in the lower portion of the nasal cavity independently of olfaction. In fact, the nasal vault could be completely occluded, while the nose still performed its role of warming, humidifying, and filtering the air through a seemingly intact airway. As a result, patients may present with a conductive or obstructive olfactory loss, yet have no other nasal symptoms.[8]

Clearly, olfactory sensitivity is completely dependent upon the ability of the nose to deliver odorant molecules to the olfactory cleft. However, due to the complex anatomy and extremely dynamic nature of the human nasal airway, it is a difficult area to study, and few reports focus on its role in the olfactory process. In vivo attempts to characterize nasal airflow have been hampered by the lack of reliable methods of observation and measurement, as well as by the tendency for any manipulations within the nose to produce a vasomotor reaction that is difficult to quantify.[9]

This led to the development of a number of in vitro techniques relying largely on plastic reconstructions of the nasal cavity that could be more easily manipulated.[9-11] While these in turn fail to account for the dynamic nature of the nasal skeleton and its mucosal lining, some valuable insights have resulted.

The nasal valve is the site of maximum resistance to inspired air, dictated largely by the angle between the septum and upper lateral cartilage, with its overall surface area.[12] Under normal conditions, air passes 60° vertically upwards through the anterior nares at a velocity of 2 to 3 meters per second. At the nasal valve, airflow changes direction from vertical to horizontal, and velocity can reach 18 meters per second. However, once passing through the valve, the cross-sectional area increases dramatically and the velocity thereby falls precipitiously. This combination of changing direction and falling velocity creates turbulent flow that allows for prolonged mucosal contact and facilitates air conditioning by the nose.[12]

During relaxed nasal breathing, airflow proceeds mainly along the level of the middle and inferior turbinate, while only a small fraction is diverted up toward the olfactory cleft. Stuiver[13] observed aluminum particles suspended in water flowing through a plastic model of the human nose and found that at normal flow rates, between 5% and 10% of the total nasal flow passed through the olfactory cleft. At higher flow rates, this increased to a maximum of 20%. More recently, Scherer et al[11] created a more anatomically correct model and found that 15% of inhaled air flows through the olfactory cleft, regardless of flow rate. As such, the best predictors of stimulus delivery to the olfactory receptors appear to be the concentration of odorant molecules within inspired air and the flow rate.[14]

However, in reality it is quite difficult to break down this process into concise variables. For example, different odorants will have different sorption characteristics that will not only influence distribution through the mucous layer overlying the olfactory receptors, but also cause a varying portion to diffuse out even before reaching the olfactory cleft.[15] In addition, as alluded to earlier, these in vitro models fail to account for mucosal changes that will alter turbulence and eddy currents that occur so characteristically in the nose and surely account in large part for the deposition of odorant molecules within the nasal vault. This raises the question as to whether even daily variations in nasal membrane function influence olfactory sensitivity.

Using citral as the odorant, Schneider and Wolf[16] performed repeated measurements of olfactory thresholds in eight subjects over a period of 3 to 36 months and correlated these with nasal membrane function. Estimates of nasal membrane function were based on a numerical scale, incorporating color, swelling, secretions, and obstruction. While these estimates were subjective and taken after olfactory testing rather than concurrently, they were nevertheless carried out by the same observer. Surprisingly, they found that olfactory acuity was best in the presence of moderate swelling, redness, and mucus secretion. Conversely, both a pale, dry, relatively shrunken membrane and a very red, very swollen membrane were associated with poor acuity. This would indicate that odorant transduction is facilitated by greater warmth and humidification within the nose provided there is not substantial airflow obstruction.

Although Schneider and Wolf[16] did not actually measure nasal resistance, their study suggests that smell functions best in the presence of moderate turbinate congestion, and therefore moderate nasal obstruction. Clearly, resistance provided by the nasal airway increases turbulence and mucosal contact time.[12] Whether there is an optimal resistance for olfaction has not been determined, nor is it clear that there is a specific resistance above which olfaction consistently cannot occur.

On the other hand, it is well documented that 80% of individuals experience cyclical changes in nasal resistance between the two nostrils.[17] Since the overall resistance remains essentially constant, normal subjects are generally unaware of this cycle, despite the fact that unilateral occlusion may be almost complete. Can the nasal cycle influence olfactory sensitivity?

Eccles et al[18] compared thresholds to both an olfactory stimulus (vanillin) and a trigeminal stimulus (L-menthol) with measures of nasal resistance based on anterior rhinomanometry. Despite an eightfold range in unilateral nasal resistance, they were able to find no effect upon unilateral threshold measures. More recently, Doty and Frye[19] determined threshold and suprathreshold measures of 2-butanone odor perception in 80 subjects and compared these with anterior rhinomanometric measures of nasal resistance in each nostril. Again, despite the changes in nasal resistance, they found no change in average threshold values.

These studies suggest that normal cyclical variations in nasal resistance are not associated with an olfactory loss, and therefore when such a loss is detected it is pathological. On the other hand, there is evidence to suggest that increasing resistance may effect perceived odor intensity.[20] The relationship between nasal resistance and sensation of airflow is quite complex, and studies have repeatedly demonstrated that rhinomanometric measures of resistance generally do not correlate with a patient's complaint of obstruction.[21] The reasons for this are speculative but may relate to trigeminal receptors that contribute to a sense of nasal patency as they help to regulate respiratory activity.[21]

Youngentob et al[20] asked 10 normal subjects to estimate the odor intensity of ethyl butyrate at four concentrations while inhaling the odorant through a monofilament mesh screen of varying mesh size. By doing so they were able to alter nasal airway resistance without creating an untoward vasomotor response. However, gas chromatographic analysis ensured that the amount of odorant reaching the nose remained the same at each given concentration despite these changes in resistance. The subjects reported decreasing odor intensity with increasing resistance, suggesting that the perceived effort of sniffing will influence the perception of odor intensity. This may be an important consideration when a complaint of olfactory loss accompanies nasal pathology, even though objective testing fails to reveal such a loss. For example, patients sniffing against an increased resistance such as a septal deviation or concha bullosa may theoretically complain of a diminished sense of smell, while olfactory threshold and suprathreshold measures remain normal.

Interestingly, this study also found that despite increasing resistance and perceived effort of sniffing, sniff duration, velocity, and volume remained constant for each subject. This supports the notion that individuals seem to develop a natural sniff technique that is optimal for their given anatomy, and sniff parameters tend to remain constant regardless of odorant type or concentration.[22]

In summary, we know that airflow patterns in the nose are complicated and difficult to measure. By design, the nose encumbers the airway with increased resistance that generates turbulent flow so that its air-conditioning tasks can be accomplished. Within this milieu, olfactory receptors must be provided with adequate stimulus to facilitate perception. As though by evolutionary design, olfaction seems to work best when the nose is performing these other functions, with moderate engorgement and mucus production. Daily variations in nasal resistance do not seem to hinder odorant access to the nasal vault, although this is complicated by individual sorption characteristics of specific odorants.

On the other hand, only about 15% of inspired air passes through the olfactory cleft, suggesting that obstruction in this area could easily cause an olfactory loss while not interfering with other nasal functions. Loss of smell may then be the only complaint without other nasal symptoms. In the presence of a complete nasal obstruction, a complete olfactory loss will undoubtedly occur. However, Schneider and Wolf's study[16] suggests that even in the presence of a widely patent airway, such as with atrophic rhinitis or excessive turbinate removal, a loss of olfactory acuity may result.

NASAL PATHOLOGY AND OLFACTORY LOSS

Any inflammatory or obstructive process in the nose can theoretically result in a loss of smell, although the extent of pathology can be quite variable. In fact, there is little information available in the literature as to specific sites or type of disease that will readily predispose to an olfactory loss. In general, such pathology can be broken down into four categories: nasal deformity causing anatomical obstruction, neoplastic disease, inflammatory disease with and without nasal polyps, and surgical trauma.

Anatomical Obstruction

The most common anatomical deformity causing nasal airflow obstruction is a deviated septum, something most general otolaryngologists routinely care for in their practice. Surprisingly, the relationship between smell loss and septal deformity has never been systematically studied. In fact, although it is often listed in textbooks as a cause of olfactory loss,[23] this is based largely on scattered anecdotal reports.

Several authors have reported cases where corrective septal surgery has improved olfactory acuity, but these patients were not tested and improvement was strictly subjective.[24,25] As discussed in the previous section, the study by Youngentob et al[20] suggests that sniffing against increased resistance will lead to a perception of decreased odor intensity, despite odorant concentration and olfactory ability remaining unchanged. This could conceivably account for perceived changes in olfactory ability associated with septal surgery and underscores the importance of objective olfactory testing.

Jafek and Hill[26] reported hyposmia in patients with severe septal deviation but did not include specifics, although it was suggested that repair must address the upper septum to open the nasal vault. Vainio-Mattila[27] evaluated a variety of anatomical and functional aspects of the nose in 162 random individuals. Two odorants were used, and the specifics of the testing method were not described, but he reported a positive correlation between increasing septal deformity and decreasing olfactory ability.

Seiden[7] recently reviewed 339 patients who presented to the University of Cincinnati Taste and Smell Center with a variety of chemosensory complaints, 289 with a primary complaint of smell loss or dysosmia. Fifty-three patients were found to have a conductive olfactory loss, but in no patient was the loss attributed to a septal deviation, suggesting that if such an etiology occurs it is quite rare. On the other hand, one could argue that such patients may be more apt to present for nasal obstruction rather than for an olfactory loss, such that any associated smell problem may go unnoticed. Indeed, the relationship between septal deformity and olfactory loss requires further study. Currently, if a patient is found to have an objectively measured loss of smell, and on physical exam is noted to have a deviated septum, further cause should be sought.

Ophir et al[28] evaluated 24 patients scheduled to have surgery for obstructing inferior turbinates and noted that 14 (58%) reported diminished olfactory acuity. However, based upon threshold testing utilizing a three-way forced-choice method and four odorants, all were considered hyposmic or anosmic. These investigators were able to demonstrate postoperative improvement, implying that hypertrophied inferior turbinates can interfere with nasal airflow to the point of impairing olfactory sensitivity. Unfortunately, the presence of any associated nasal conditions was not described or denied.

Neoplastic Disease

Any neoplastic process arising in the nose or paranasal sinuses that completely obstructs the olfactory cleft can result in a conductive olfactory loss. Such lesions are not common, but rarely do they present with olfactory loss as the primary symptom. For example, in 289 patients presenting to a university-based taste and smell clinic with a primary complaint of olfactory dysfunction, only one patient had such a neoplasm—specifically, a recurrent inverting papilloma.[7] Presenting symptoms will depend upon the size and location of the lesion, but typically will include nasal obstruction or epistaxis.

The incidence of chemosensory dysfunction resulting from these various lesions is not known. However, the loss in such cases may not only be conductive, but also sensorineural as a result of direct invasion of the olfactory neuroepithelium. A complete discussion of tumors of the nose and paranasal sinuses is beyond the scope of this chapter.

Rhinitis, Sinusitis, and Nasal Polyps

Inflammatory lesions of the nose and paranasal sinuses account for most cases of conductive olfactory loss. Of 53 patients presenting with a primary complaint of olfactory loss in whom the loss was found to be conductive, Seiden[7] found that 50 had rhinitis, sinusitis, or nasal polyps.[7] Two of the remaining three patients had a smell loss from postsurgical trauma, and one patient had an inverting papilloma.

Breaking this group down further, it was determined that 21 patients (40%) had nasal polyps, 19 patients (36%) had chronic sinusitis without intranasal polyps, 8 patients (15%) had allergic rhinitis, and 1 patient (2%) had atrophic rhinitis (Table 5–1).[7] One patient presented with dysosmia secondary to chronic sphenoid sinusitis but had normal olfactory acuity.

As might be expected, those patients with intranasal polyps generally demonstrated a more severe smell loss, with a mean score on the University of Pennsylvania Smell Identification Test (UPSIT) of 15.7, as compared to that group with chronic sinusitis and no polyps (19.8). Although both groups were on average anosmic, this difference was found to be statistically significant.[7] Nine of these patients had polyps strictly within the nasal vault, including the superior meatus and sphenoethmoidal recess, while the middle meatus was clear (Fig. 5–1); five patients had polyps confined to the middle meatus (Fig. 5–2); and seven patients had more diffuse disease, with polyps present in both the middle meatus and nasal vault (Table 5–1). Patients with only allergic rhinitis were more likely to be hyposmic, with a mean UPSIT score of 25.6.

Table 5–1   Conductive Olfactory Loss (N = 53)

	N	UPSIT (N)
Intranasal polyps	21	15.7
Nasal vault	9	18.4
Middle meatus	5	14.0
Diffuse	7	13.3
Chronic sinusitis	20	20.4
Allergic rhinitis	8	24.4
Inverting papilloma	1	10.0
Previous surgery	2	24.0

UPSIT = University of Pennsylvania Smell Identification Test. From Seiden.[7]

It bears repeating that these patients presented with olfactory loss as their primary symptom. The true incidence of olfactory dysfunction associated with inflammatory conditions of the nose is difficult to determine, as many such patients will present for other reasons. Estimates indicate that 10% to 15% of the general population suffer from allergic rhinitis,[29] making inhalant allergy one of the most common causes of nasal symptoms. Yet only two studies attempt to look at the incidence with which olfactory loss occurs in association with allergic rhinitis.

Seiden et al[30] administered the UPSIT to 37 patients with active allergic rhinitis verified by skin testing and found that 13 (35%) had a measurable loss of smell. Seven patients had associated nasal polyps, and their smell function was significantly worse than in those without polyps. Nevertheless, 6 patients (16%) without polyps had an objective olfactory loss associated with allergic rhinitis. Unfortunately, possibly concurrent sinus disease was not discussed.

Cowart et al[29] assessed olfactory thresholds in 91 patients with active allergic rhinitis based on skin testing and 80 nonatopic controls. In the former group, 23.1% demonstrated a clinically significant loss of smell, compared with only 2.5% in the control group. Twenty-eight patients had associated sinusitis, nasal polyps, or both, and 12 (42.9%) were noted to be hyposmic. On the other hand, of 63 patients manifesting allergic rhinitis only, 9 (14.3%) were hyposmic. Although nasal resistance as measured by rhinomanometry was significantly higher among those with allergic rhinitis as a group than among controls, the investigators found no correlation between resistance and olfactory threshold.

Therefore, it seems clear that allergic rhinitis alone can with substantial frequency cause a measurable olfactory loss. Experience in turn indicates that more extensive nasal disease associated with infectious sinusitis or nasal polyps has a more pronounced effect upon olfactory function.[7] However, it remains curious why some patients with apparently extensive disease may report no loss of smell, while others with seemingly mild disease cannot smell at all. Clearly this must relate to obstruction or disruption of airflow to the nasal vault.

The incidence of smell loss in patients presenting with nasal polyps ranges from 11% to as high as 48%, although generally these reports do not rely upon currently acceptable olfactory testing methods.[31] Hosemann et al[32] reviewed 111 patients with

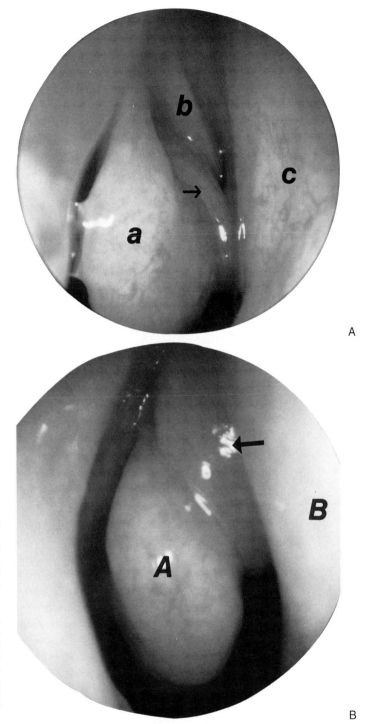

Figure 5–1. **A**: Examination of the right nasal cavity utilizing rigid nasal endoscopy. A polyp can be seen protruding from the superior meatus (arrow), while the middle meatus is clear. a = middle turbinate, b = superior turbinate, c = septum. **B**: Another example of a polyp (arrow) seen filling the nasal vault, while the middle meatus is clear. This is a right nasal cavity, visualized with a rigid endoscope. A = middle turbinate, B = septum.

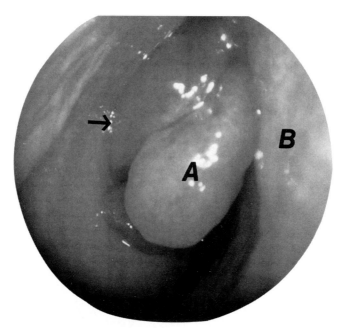

Figure 5–2.   Examination of the right nasal cavity in this patient demonstrates polypoid disease within the right middle meatus (arrow). A = middle turbinate, B = septum.

chronic polypoid ethmoiditis and found the presenting symptoms to be nasal obstruction in 27%, headache and facial pain in 19%, and olfactory loss in 17%. However, by performing both threshold and identification testing in these patients, they found in fact that 65% had a measurable smell loss.

Surgical Trauma

Despite the frequency with which surgery is performed on the nose to improve the nasal airway or its external appearance, rarely is the affect upon olfaction considered. Theoretically, olfactory loss could be sensorineural as a result of direct injury to the cribriform plate and its delicate neuroepithelium, or it may be conductive, due to surgical scarring obstructing the nasal vault or the development of chronic hyperplastic sinusitis following surgical failure. It is not customary to measure preoperative olfactory function unless the patient presents specifically with a chemosensory complaint, and this makes it difficult to substantiate a postoperative deficit.

Champion[33] had noted anecdotally that several patients reported loss of smell following rhinoplasty and therefore interviewed 200 patients following the operation. He found that 10% complained of postoperative olfactory loss lasting from 6 to 18 months, but all eventually did recover, except for one patient. However, he relied solely on subjective reports.

Goldwyn and Shore[24] utilized coffee, oil of peppermint, and oil of clove to test a series of 97 patients undergoing various combinations of rhinoplasty and submucous resection. They reported no evidence of olfactory loss at 2 months.

However, as pointed out by Doty and Frye,[19] the crude method of olfactory testing described would have been insensitive to anything but extreme alterations in smell ability. In addition, the failure to use sound psychophysical principles could be further criticized.[19]

Stevens and Stevens[34] prospectively evaluated 100 patients undergoing nasal surgery, conducting olfactory testing 3 to 4 weeks postoperatively. Eight were found to have a decrease in olfactory sensitivity, with 1 patient becoming completely anosmic after a septorhinoplasty. This would suggest that the risk of surgery to the sense of smell is quite low. However, these patients underwent a variety of nasal procedures, making comparison more difficult. In addition, the authors used a modification of the Elsberg blast technique to measure olfaction, a method that has been largely condemned due to its introduction of an unnatural stimulus pulse into the nose.[19,35]

Wigand and Hosemann[36] reviewed 600 patients who had undergone endoscopic ethmoidectomies and found the incidence of olfactory loss to be approximately 3%. They attributed this complication to inadequate sparing of the medial plate of the middle turbinate. After modifying their technique, the same group reported a smaller series of 111 patients and found that 2% suffered a worsening of their ability to smell.[32] Although there has been some suggestion in the literature that olfactory fila extend down into the middle turbinate,[37] a recent histopathological study found no such evidence.[38] Accessibility of the olfactory cleft appears to be more important, and therefore scarring in this area needs to be avoided.

In an attempt to accurately define the surgical risk to olfaction, Kimmelman[39] evaluated 93 patients undergoing various nasal procedures. The UPSIT was utilized to assess both preoperative and postoperative olfactory function. Sixty-one patients (66%) improved or maintained their preoperative scores, and whereas 32 patients (34%) had a decline in score, this decline was not statistically significant. Only 1 patient bacame anosmic. Therefore, Kimmelman concluded that the risk of becoming anosmic after nasal surgery is approximately 1.1%, regardless of the type of operation, the anesthetic used, or the age and sex of the patient.

THE DIAGNOSIS OF OLFACTORY LOSS SECONDARY TO NASAL AND SINUS PATHOLOGY

In some cases the conductive nature of an olfactory loss may be quite apparent. However, as alluded to previously, the history and physical exam can be misleading, and olfactory dysfunction inadvertently assumed to be sensorineural. By the same token, the discovery of nasal or sinus pathology in a patient presenting with olfactory loss does not necessarily prove cause and effect. A number of diagnostic points are helpful toward making a correct diagnosis. Therefore, it is imperative that any patient presenting with a loss of smell undergo a thorough evaluation.

History

The history must attempt to establish characteristic features of associated nasal and sinus inflammation, as well as rule out other potential causes of olfactory dysfunction, eg, trauma, viral infection, and toxic exposure (see chapters 6 to 8). Causing considerable confusion is the fact that symptoms of a viral upper respiratory infection may precede or parallel those of rhinitis or sinusitis, and patients with chronic sinus disease will often be prone to recurrent bouts of acute infection that may mask associated viral illness. The distinction may therefore prove to be quite difficult.

It has been previously reported that nasal and sinus pathology generally produces a severe loss of smell, such that most patients will be anosmic when tested.[4,40] In a recent review, it was found that of 53 patients with a conductive olfactory loss, 30 (57%) were anosmic, with a mean UPSIT score of 11.6, while 21 (40%) were hyposmic, with a mean UPSIT score of 27.4.[7] However, this difference was not statistically significant, and therefore not likely to be of diagnostic benefit. A history of chronic sinusitis was described by 59% of patients, as was a history of inhalant allergies, whereas only 23% had a history of asthma. It is of note that Scott et al[40] have found a significant incidence of inhalant allergies (35%), chronic nasal symptoms (42%), and a past history of nasal and sinus disease (25%) in patients diagnosed with viral olfactory loss.

Intuitively, it would seem that nasal obstruction would be a common symptom in patients with a conductive olfactory loss. However, in Seiden's series only 32% of 53 patients complained of obstruction, demonstrating that the nasal airway may remain patent despite complete obstruction of the nasal vault.[7] Such patients are apt to present primarily for their loss of smell, and the clinician must not be confused into thinking the loss is not likely to be conductive.

A history suggesting transient fluctuations in olfactory sensitivity is consistent with a conductive etiology because it implies an intact olfactory neuroepithelium and does not occur in the case of a sensorineural loss. Although recovery of olfactory function has been described in cases of viral-induced loss, this tends to be subtle and very gradual, and not prone to fluctuation.[5] Such fluctuation might be noted after any activity that facilitates mucosal decongestion in the nose, such as exercise, showering, or perhaps with the use of certain medications.[31] A history of fluctuation should therefore encourage a search for underlying sinus pathology. However, again in Seiden's series, only 45% of patients with a conductive loss experienced fluctuation.[7] Therefore, this factor cannot be relied upon to make the diagnosis.

Complaints of dysosmia, whether a persistent phantom odor (phantosmia) or distorted odor (parosmia), are not common in patients with a conductive olfactory loss. An incidence of 34% has been reported in these patients, as compared to 65% in patients with a postviral olfactory loss.[7,41] Therefore, complaints of dysosmia may suggest a viral etiology. However, a foul odor may occasionally arise from underlying purulent sinusitis, producing complaints of either phantosmia or parosmia, and in such cases may occasionally be detected by other family members.

When one takes a history, questions need to elicit any symptoms that would be consistent with underlying nasal or sinus disease. Complaints of nasal obstruction and discharge and a history of chronic sinusitis or allergies should lead the clinician to suspect such disease as the probable cause and lead to appropriate steps to verify the diagnosis. On the other hand, the absence of such symptoms does not necessarily rule out a conductive etiology, as olfactory loss may be its only manifestation.

Chemosensory Testing

Some form of chemosensory testing is essential in any patient presenting with a complaint of olfactory loss. However, to be meaningful, it must be based upon sound psychophysical testing principles and be reproducible, yet practical in a clinical setting. Many earlier studies that relied on subjective reports or the ability of patients to identify one or two random odorants are of questionable validity and difficult to interpret. Fortunately, several tests have now been developed and standardized on normal populations and are readily accessible to most clinicians. The reader is referred to chapter 4 for further information.

For olfactory testing to establish the conductive nature of an olfactory loss, a given test would have to be able to bypass the nose to measure sensorineural function. While a number of neurophysiological approaches have been explored, such as measuring summated generator potentials directly from the neuro-epithelium or recording olfactory-evoked potentials, these techniques remain largely experimental.[42,43] Currently acceptable techniques for detecting olfactory loss can reliably determine the degree of loss but cannot distinguish between a conductive and sensorineural etiology.

Physical Examination

Based upon the previous discussion regarding nasal physiology and olfaction, findings of moderate turbinate congestion and septal deviation are unlikely to be associated with an objective olfactory loss. On the other hand, pronounced rhinitis, nasal polyps, or evidence of sinus inflammation or infection would implicate a conductive etiology.

Most otolaryngologists routinely perform anterior rhinoscopy to evaluate the nose, usually before and after nasal decongestion. While this is very important in patients presenting with olfactory loss, utilizing a nasal speculum exclusively will in some cases lead to an inadequate exam and an inappropriate diagnosis. Messerklinger[44] and Kennedy et al[45] have stressed the advantages of incorporating rigid endoscopy for nasal diagnosis, but although it is becoming more prevalent, it still is not a routine part of every head and neck examination. When patients present with olfactory loss or dysosmia, despite the fact that other nasal or sinus symptoms may be absent, nasal endoscopy is extremely valuable. While a patient's history may suggest a possible viral etiology or lead the examiner to suspect some other sensorineural pathology, nasal endoscopy can detect disease within the nasal vault or ostiomeatal complex that otherwise might be missed.[7] Such findings would then lead to appropriate steps to verify the conductive nature of the olfactory loss, at which point therapy could be instituted to restore the sense of smell.

In a recent review of patients presenting with a conductive olfactory loss, Seiden[7] compared anterior rhinoscopy with nasal endoscopy in terms of the ability of each to detect the presence of nasal pathology. In 24 of 50 cases (48%) anterior rhinoscopy was unremarkable, whereas nasal endoscopy revealed underlying pathology. The most common endoscopic diagnosis in these 24 patients was ostiomeatal pathology without intranasal polyps, noted in 11 patients (Fig. 5–3),

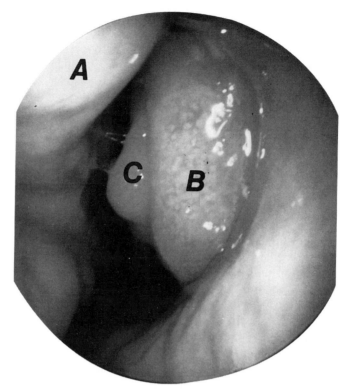

Figure 5–3. Rigid nasal endoscopic exam of a left nasal cavity reveals evidence of ostiomeatal disease, with edema and thickening of the uncinate process (B) with edema also noted of the ethmoidal bullae (C). A = middle turbinate.

Table 5–2 Nasal Endoscopic Findings With Negative
Anterior Rhinoscopy (N = 24/50: 48%)

Ostiomeatal pathology	11
Polyps nasal vault	8
Polyps middle meatus	2
Polyps nasal vault and middle meatus	1
Sphenoid sinusitis	1
Postsurgical hyperplastic sinusitis	1

From Seiden.[7]

followed by polyps within the nasal vault, noted in 8 patients (Fig. 5–4, Table 5–2). Two patients had polyps within the middle meatus, and the remaining three patients had polyps within the nasal vault and middle meatus, sphenoid sinusitis, and post-surgical hyperplastic sinusitis, respectively. In only four cases (8%) was the diagnosis unclear after both anterior rhinoscopy and nasal endoscopy, and therefore made by computed tomography scan.

Although Davidson (cited in Hill and Jafek)[46] has reported endoscopically visualizing the olfactory neuroepithelium and found this to be of some diagnostic use, this is not feasible in most cases, particularly those with nasal pathology.

However, endoscopic examination affords better visualization of the nasal vault, the ostiomeatal complex, and sphenoethmoidal recess and greatly enhances the examiner's ability to detect underlying sinus disease. It should be considered in all patients presenting with an olfactory loss, even if the findings on anterior rhinoscopy are unremarkable and whether or not the history suggests underlying sinus disease.

Use of Topical and Systemic Steroid Medications

Given that a conductive olfactory loss occurs due to inflammatory obstruction of the olfactory cleft, steroids as potent anti-inflammatory agents might be expected to reverse such a loss. This would not only provide some therapeutic benefit but would offer a method to verify that the loss is indeed reversible. Such reversibility would serve to exclude a sensorineural etiology, including a viral-induced loss, which has not been found to respond to therapeutic intervention.[6]

Several limited studies have looked at the effects of steroid administration upon olfactory loss. Hotchkiss[47] reported a series of 30 patients with massive nasal polyposis and associated anosmia. After prednisone starting at 30 mg/d tapered over 7 days, subjective restoration of smell was described in proportion to the shrinkage of polypoid disease.

Fein et al[48] selected 18 patients complaining of olfactory loss and allergic rhinitis, although most had other nasal pathology, including nasal polyps. Various combinations of treatment were untilized, such as immunotherapy and surgery, but these authors reported the most significant improvement in association with the use of systemic steroids. Nevertheless, the uncontrolled nature of this study makes it difficult to draw conclusions.

More recently, Goodspeed et al[49] tested olfactory function in 20 anosmic and hyposmic patients before and after a one-week course of systemic steroids. Ten patients had nasal and sinus disease, 4 had a postviral olfactory loss, and 6 were considered idiopathic. Only 6 patients responded, all of whom had a conductive loss.

In a preliminary study, Scott et al[50] tried to determine the effectiveness of flunisolide spray on conductive olfactory loss. Seven patients with perennial rhinitis and nasal polyposis were placed on 2 weeks of antibiotics, at which point flunisolide and a nasal decongestant spray were introduced. Moffat's position, which is a head down and forward position that was first described for the administration of topical nasal anesthesia, and more recently in the treatment of nasal polyps, was used to administer the steroid spray[51] (Fig. 5–5) in an attempt to facilitate distribution of the spray to the nasal vault. After 2 weeks of steroid therapy, these authors reported a return of smell function to the midhyposmic range in 5 of the seven patients.

In a series of 31 patients with a conductive olfactory loss receiving systemic steroids, Seiden[7] reported that 26 (84%) experienced an improvement in olfactory acuity, albeit temporary. On the other hand, only 11 of 47 patients (23%) given topical steroid preparations noted a similar improvement, a difference that was found to be statistically significant.

Whether steroids effect this reversal by a decrease in mucosal edema or some other anti-inflammatory mechanism is unclear. It has been suggested that some

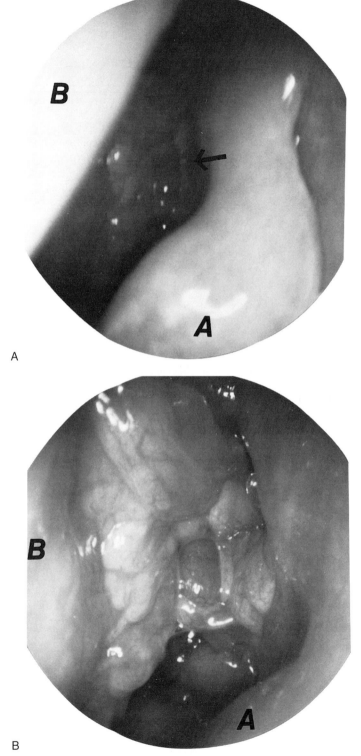

A

B

Figure 5–4. **A**: Rigid nasal endoscopic examination of a left nasal cavity suggests disease high in the nose, while the middle meatus seems clear. Anterior rhinoscopy was unremarkable. A = middle turbinate, B = septum. **B**: A closer look at the upper nasal cavity reveals polypoid disease filling the nasal vault, obstructing the olfactory cleft. A = upper middle turbinate, B = upper septum.

adverse cellular event may be influenced by the use of steroids; however, concrete evidence is lacking.[52] In any event, systemic steroids may help to verify the conductive nature of an olfactory loss, while failure to respond to an intranasal steroid spray is inconclusive.

Radiological Evaluation

It is generally accepted that plain radiographs are relatively insensitive and nonspecific, particularly in terms of evaluating the ethmoid sinuses and ostiomeatal complex. Therefore, they are for the most part inadequate for evaluating patients with an olfactory loss.[53]

On the other hand, computed tomography is now regarded as the imaging procedure of choice for the paranasal sinuses, able to capture both soft tissue disease as well as bony changes with appropriate window adjustments.[54] Thin sections in the coronal plane are usually ideal, demonstrating the ostiomeatal complex, superior nasal cavity, and olfactory cleft (Fig. 5–6).

When a patient presents with an olfactory loss, the history and physical exam are essential to determine whether the loss is likely to be obstructive. Scott et al[40] correctly point out that a negative head and neck exam does not necessarily rule out nasal and sinus disease as a cause of olfactory loss. They reported positive computed

Figure 5–5.   Head down and forward position known as Moffat's position, used to facilitate distribution of topical steroid sprays to the nasal vault.

tomography scans in 24% of patients with normal exams; however, the details of that exam are not described. Anterior rhinoscopy coupled with nasal endoscopy should be able to detect underlying pathology in better than 90% of cases.[7] Therefore, although high-resolution computed tomography is the most definitive way to rule out a conductive etiology, it may not be necessary in every case. If the history is strongly suggestive, or subtle findings on endoscopy suggestive, then further x-ray study is warranted. Computed tomography scanning is also helpful in delineating the full extent of disease, and if surgical intervention is planned.

If after a thorough history and examination the etiology remains idiopathic, or if intracranial pathology is suspected, then magnetic resonance imaging is the procedure of choice.[55] However, the lack of bony detail and the possibility of false-positive findings of ethmoid disease limit its usefulness in the evaluation of the paranasal sinuses and nasal cavity.[56]

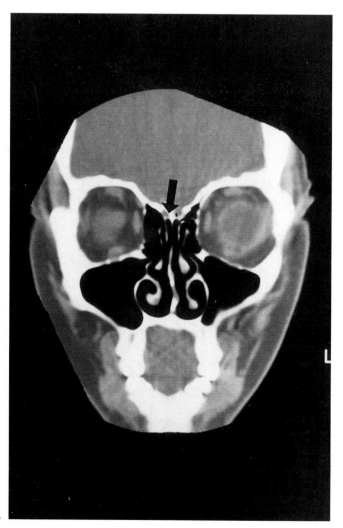

A

Figure 5–6. A: Computed tomography (CT) scan with coronal view through the anterior ethmoid sinus clearly displays the ostiomeatal complex, with no evidence of disease. The olfactory cleft is patent (arrow). **B**: Coronal CT scan demonstrates inflammatory ethmoid disease along with obstruction of the olfactory cleft (arrow).

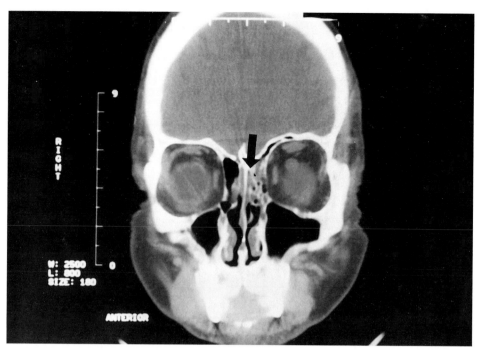

B

Figure 5–6. *Continued*

REFERENCES

1. Kennedy DW: First-line management of sinusitis: A national problem? Overview. Otolaryngol Head Neck Surg 103:847–854, 1990.
2. Goodspeed RB, Gent JF, Catalanotto FA: Chemosensory dysfunction: Clinical evaluation results from a taste and smell clinic. Postgrad Med 81:251–260, 1987.
3. Smith DV, Frank RA, Pensak ML, Seiden AM: Characteristics of chemosensory patients and a comparison of olfactory assessment procedures. Chem Senses 12:698, 1987.
4. Deems DA, Doty RL, Settle RG, et al: Smell and taste disorders, a study of 750 patients from the University of Pennsylvania Smell and Taste Center. Arch Otolaryngol Head Neck Surg 117:519, 1991.
5. Duncan HJ, Seiden AM: Long-term follow-up of olfactory loss secondary to head trauma and upper respiratory tract infection. Arch Otolaryngol Head Neck Surg 121:1183–1187, 1995.
6. Mott AE, Leopold DA: Disorders in taste and smell. Med Clin North Am 75:1321–1353, 1991.
7. Seiden AM: The diagnostic evaluation of conductive olfactory loss. Presented as a candidate's thesis to the American Rhinologic, Otologic, and Laryngologic Society, 1995.
8. Seiden AM, Smith DV: Endoscopic intranasal surgery as an approach to restoring olfactory function. Chem Senses 13:736, 1988.
9. Hornung DE, Leopold DA, Youngentob SL, et al: Airflow patterns in a human nasal model. Arch Otolaryngol Head Neck Surg 113:169–172, 1987.
10. Girardin M, Bilgen E, Arbour P: Experimental study of velocity fields in a human nasal fossa by laser anemometry. Ann Otol Rhinol Laryngol 92:231–236, 1983.
11. Scherer PW, Hahn II, Mozell MM: The biophysics of nasal airflow. Otolaryngol Clin North Am 22:265–278,1989.
12. Knops JL, McCaffrey TV, Kern EB: Physiology. Otolaryngol Clin North Am 26:517–534, 1993.
13. Stuiver M: Biophysics of the sense of smell. Doctoral thesis, Rijke University, Groningen, The Netherlands, 1958.
14. Mozell MM, Sheehe PR, Swiek SW, Kurtz DB, Hornung DE: A parametric study of the stimulation variables affecting the magnitude of the olfactory nerve response. J Gen Physiol 83:233–267, 1984.

15. Mozell MM, Kent PF, Scherer PW, Hornung DE, Murphy SJ: Nasal airflow. In: Getchell TV, Doty RL, Bartoshuk LM, Snow JB (eds): Smell and Taste in Health and Disease. New York: Raven Press, 1991.

16. Schneider RA, Wolf S: Relation of olfactory acuity to nasal membrane function. J Appl Physiol 15:914–920, 1960.

17. Principato JJ, Ozenberger JM: Cyclical changes in nasal resistance. Arch Otolaryngol 91:71–77, 1970.

18. Eccles E, Jawad MSM, Morris S: Olfactory and trigeminal thresholds and nasal resistance to airflow. Acta Otolaryngol 108:268–273, 1989.

19. Doty RL, Frye R: Influence of nasal obstruction on smell function. Otolaryngol Clin North Am 22:397–411, 1989.

20. Youngentob SL, Stern NM, Mozell MM, Leopold DA, Hornung DE: Effect of airway resistance on perceived odor intensity. Am J Otolaryngol 7:187–193, 1986.

21. Jones AS, Willatt DJ, Durham M: Nasal airflow: Resistance and sensation. J Laryngol Otolaryngol 103:909–911, 1989.

22. Laing DG: Characterisation of human behaviour during odour perception. Perception 11:221–230, 1982.

23. Doty RL, Bartoshuk LM, Snow JB: Causes of olfactory and gustatory disorders. In: Getchell TV, Doty RL, Bartoshuk LM, Snow JB (eds): Smell and Taste in Health and Disease. New York: Raven Press, 1991.

24. Goldwyn RM, Shore S: The effects of submucous resection and rhinoplasty on the sense of smell. Plast Reconstr Surg 41:427–432, 1968.

25. Kittel G, Waller G: Smell improving effect of Cottle's septum operation. Z Laryngol Rhinol Otol Ihre Grenzgegiele 52:280–284, 1973.

26. Jafek BW, Hill DP: Surgical management of chemosensory disorders. ENT J 68:398–404, 1989.

27. Vainio-Mattila J: Correlations of nasal symptoms and signs in random sampling study. Acta Otolaryngol Suppl 318:5–48, 1974.

28. Ophir D, Gross-Isseroff R, Lancet D, Marshak G: Changes in olfactory acuity induced by total inferior turbinectomy. Arch Otolaryngol Head Neck Surg 112:195–197, 1986.

29. Cowart BJ, Flynn-Rodden K, McGeady SJ, Lowry LD: Hyposmia in allergic rhinitis. J Allergy Clin Immunol 91:747–751, 1993.

30. Seiden AM, Litwin A, Smith DV: Olfactory defictis in allergic rhinitis. Chem Senses 14:746–747, 1989.

31. Mott AE: Topical corticosteroid therapy for nasal polyposis. In: Getchell TV, Doty RL, Bartoshuk LM, Snow JB (eds): Smell and Taste in Health and Disease. New York: Raven Press, 1991.

32. Hosemann W, Goertzen W, Wohlleben R, Wolf S, Wigand ME: Olfaction after endoscopic endonasal ethmoidectomy. Am J Rhinol 7:11–15, 1993.

33. Champion R: Anosmia associated with corrective rhinoplasty. Br J Plast Surg 19:182–185, 1966.

34. Stevens CN, Stevens MH: Quantitative effects of nasal surgery on olfaction. Am J Otolaryngol 6:264–267, 1985.

35. Elsberg C, Levy I: The sense of smell: I. A new and simple method of quantitative olfactometry. Bull Neurol Inst NY 4:5–19, 1935.

36. Wigand ME, Hosemann W: Microsurgical treatment of recurrent nasal polyposis. Rhinology 8:25–30, 1989.

37. Wigand ME: Endoscopic Surgery of the Paranasal Sinuses and Anterior Skull Base. New York: Thieme Medical Publishers, 1990.

38. Biedlingmaier JF, Whelan PJ, Zoarski G, Rothman M: Histopathology and CT analysis of partially resected middle turbinates. Laryngoscope 106:102–104, 1996.

39. Kimmelman CP: The risk to olfaction from nasal surgery. Laryngoscope 104:981–988, 1994.

40. Scott AE, Cain WS, Leonard G: Nasal/sinus disease and olfactory loss at the Connecticut Chemosensory Clinical Research Center. Chem Senses 14:745, 1989.

41. Duncan HJ, Seiden AM, Paik SI, Smith DV: Differences among patients with smell impairment resulting from head trauma, nasal disease, or prior upper respiratory infection. Chem Senses 16:517, 1991.

42. Ottoson D: The electro-olfactogram: A review of studies on the receptor potential of the olfactory organ. In: Beidler LM (ed): Handbook of Sensory Physiology, Olfaction, vol. 4. New York: Springer-Verlag, 1971.

43. Kobal G, Hummel T: Olfactory evoked potentials in humans. In: Getchell TV, Doty RL, Bartoshuk LM, Snow JB (eds): Smell and Taste in Health and Disease. New York: Raven Press, 1991.

44. Messerklinger W: Endoscopy of the Nose. Baltimore: Urban and Schwarzenberg, 1978.

45. Kennedy DW, Zinreich SJ, Rosenbaum AE, Johns ME: Functional endoscopic sinus surgery: Theory and diagnosic evaluation. Arch Otolaryngol 111:576–582, 1985.
46. Hill DP, Jafek BW: Initial otolaryngologic assessment of patients with taste and smell disorders. ENT J 68:362–370, 1989.
47. Hotchkiss WT: Influence of prednisone on nasal polyposis with anosmia. Arch Otolaryngol 64: 478–479, 1956.
48. Fein BT, Kamin PB, Fein NN: The loss of sense of smell in nasal allergy. Ann Allergy 24:278–283, 1966.
49. Goodspeed RB, Gent JF, Catalanotto FA, Cain WS, Zagraniski PT: Corticosteroids in olfactory dysfunction. In: Meiselman HL, Rivlin RS (eds): Clinical Measurement of Taste and Smell. New York: Macmillan, 1986.
50. Scott A, Cain WS, Clavet G: Topical corticosteroids can alleviate olfactory dysfunction. Chem Sense 13:735, 1988.
51. Chalton R, Mackay, I. Wilson R, Cole P: Double blind, placebo controlled trail of betamethasone nasal drops for nasal polyposis. Br Med J 291:788, 1985.
52. Jafek BW, Moran DT, Eller PM, Rowley JC, Jafek TB: Steroid-dependent anosmia. Arch Otolaryngol Head Neck Surg 113:547–549, 1987.
53. Goodspeed RB, Gent JF, Leonard G, Catalanotto FA: The prevalence of abnormal paranasal sinus x-rays in patients with olfactory disorders. Connecticut Med 51:1–3, 1987.
54. Zinreich SJ, Kennedy DW, Rosenbaum AE, Gayler BW, Kumar AJ, Stammberger H: CT of nasal cavity and paranasal sinuses: Imaging requirements for functional endoscopic sinus surgery. J Radiol 163:769–775, 1987.
55. Li C, Yousem DM, Doty RL, Kennedy DW: Neuroimaging in patients with olfactory dysfunction. Am J Rhinol 162:411–418, 1994.
56. Zinreich SJ, Kennedy DW, Kumar AJ, Rosenbaum AE, Arrington JA, Johns ME: MR imaging of normal nasal cycle: Comparison with sinus pathology. J Comput Assist Tomogr 12:1014–1019, 1988.

6

Postviral Olfactory Loss

HEATHER J. DUNCAN, Ph.D.

Olfactory dysfunction following a viral-like upper respiratory infection (URI) was first described in several patients in 1975.[1] This sensory loss is one of the most common identifiable causes of olfactory dysfunction. This etiology alone accounts for 20% to 25% of patients presenting to the University of Cincinnati Taste and Smell Center with olfactory complaints, a frequency similar to that reported at other chemosensory dysfunction centers.[2–4]

PATIENT CHARACTERISTICS

Temporal contiguity between the "viral" episode and recognition of olfactory impairment is the major feature in the diagnosis of prior URI as the cause of olfactory impairment. To date, no simple diagnostic tests are available to identify these patients, even if patients could be tested during the episode of illness. While patients report these "colds" as severe, they do not always seek medical help, nor do they use the same over-the-counter medicines to find relief. Thus, determining commonalities among patients who have suffered anosmia or hyposmia following a "viral" infection has received a great deal of attention in the past few years and has increased the precision with which this diagnosis can be applied to an individual patient.

   The demographics of patients presenting to smell and taste centers with URI-induced smell loss present a different profile than in other etiological categories. In particular, these patients are older and more likely to be women.[2–4] This gender and age bias dictates that sensory testing must be chosen that will provide comparisons to normative data for olfactory function according to gender and age (such as the University of Pennsylvania Smell Identification Test [UPSIT][5]). Even though a patient may report a sudden onset of symptoms following an URI, if sensory testing shows olfactory sensitivity to be in the range considered normal for age- and gender-matched controls, it is not possible to rule out the contributions that aging may have made to decreased olfactory function, and thus the prognosis for recovery may be compromised.

Patients reporting loss of olfactory function following URI may or may not suffer from other medical conditions, but they usually do not have a history of nasal/sinus disease and, in fact, must have a negative otolaryngological exam to be considered a URI patient.[6]

SENSORY EVALUATION

Typically, olfactory loss is measured by threshold and/or odor identification tasks, such as the UPSIT, and patients are characterized as hyposmic or anosmic. The measured degree of sensory loss for URI patients is, on average, less severe than for head trauma or nasal/sinus disease patients.[7-9] Unpleasant parosmias or phantosmias are reported frequently; the proportion of patients with URIs reporting either type of dysosmia is higher (65%) than the proportion of patients with these complaints in nasal/sinus disease or head trauma (34% and 35%, respectively[10]). Although other studies do not find this increased incidence of dysosmia in URI patients,[3] it is consistent with the higher level of olfactory function seen in these patients compared to other etiologies.[3]

Approximately one third of URI patients report dysgeusias, either altered taste perceptions or phantom taste sensations. In many cases, the initial complaint may be a decreased or absent sense of taste. However, with a few exceptions,[1] measurable taste loss is rare[3,11] and the complaint of taste loss reflects a loss of flavor sensitivity.[12]

PATHOLOGY

The mechanism for sensory loss after upper respiratory infection is not well understood, but it is well known that viral invasion of the central nervous system can occur through the route of the olfactory nerves.[13] The effects of various viral infiltrations differ, depending upon whether the virus itself destroys olfactory receptor cells or whether cellular damage is caused by the immune response to the virus. Olfactory system damage can be at the level of the epithelium or can occur in central processing pathways, ie, the olfactory bulb, leading to retrograde degeneration of the olfactory receptor cells, perhaps with incomplete regeneration.[11]

Central Effects

The effects of viral infiltration on the central olfactory system can be modeled in animal studies, although little is known about the central effects of viral infection on olfactory sensitivity in the human. In studies of vesicular stomatitis virus in rats, for example, reductions in serotonin levels were found in several brain areas, including the olfactory bulb and hippocampus, following intranasal inoculation.[14,15] These effects persisted until at least 18 months of age and were accompanied by several behavioral deficits, as seen in altered activity levels and altered behavior in the

Morris water maze, which is a measure of general cognitive deficits. No measures of olfactory function were undertaken.[15]

Herpes simplex virus (type 1, HSV-1) has also been shown to affect central olfactory pathways following intranasal innoculation.[16] Herpes simplex virus type 1 can produce encephalitis with accompanying behavioral deficits.[13] Presumably these behavioral deficits are due to focal sites of necrosis caused by the viral invasion, although dopamine function is altered and may play a role in behavioral abnormalities.[17] No systematic examinations of olfactory function after HSV-1 infection in either humans or animals have been undertaken.

Studies of the coronavirus mouse hepatitis virus (MHV) have also shown the ability of the virus to be transported transneuronally into the central nervous system (CNS) via olfactory and nasal trigeminal pathways.[18,19] Since the MHV may replicate in the nasal epithelium,[18] studies of its effect on olfactory function would be useful but are yet to be done.

Inhalation of an aerosolized rabies virus has also been reported to produce neuronal destruction consistent with encephalitis and evidence of virus within the olfactory bulb, in one patient accidently exposed in a laboratory.[20] Although the olfactory bulbs of this patient were examined ultrastructurally, no determination of olfactory function or examination of olfactory epithelium was reported; the intranasal route was verified by evaluation of the laboratory conditions.

It is to be hoped that future studies of viral invasion of the CNS via olfactory pathways will include assessment of olfactory function. This is difficult in animal models, but studies of both parainfluenza virus and canine distemper in dogs have included a simple behavioral measure of olfactory sensitivity.[21,22] In animals naturally and experimentally exposed to parainfluenza virus, olfactory thresholds were elevated during the infection but returned to normal following resolution of the disease. This finding, coupled with normal olfactory receptor function (as measured by electro-olfactography) and changes in respiratory epithelium, caused the investigators to conclude that hyposmia was a result of abnormal airflow patterns of normal olfactory stimuli, induced by inflammation of the respiratory epithelium. Thus, even this study failed to produce a model of the olfactory dysfunction seen in humans after URI.[21]

Canine distemper virus, on the other hand, produced an apparently permanent anosmia, as behavioral measures of olfactory sensitivity as well as electro-olfactograms showed dysfunction even 6 months after cessation of clinical signs of disease. Nasal mucosa taken from animals with active disease showed, in addition to edema, lesions in both respiratory and olfactory regions, with more severe damage (atrophy and erosion) in the olfactory epithelium. This study provides a link between viral infection, olfactory dysfunction, and epithelial damage due to virus.[22]

Peripheral Effects

Either central degeneration of olfactory pathways or destruction of the olfactory epithelium could result in a relatively long-lasting hyposmia or anosmia, and at present these are not clinically distinguishable. The olfactory epithelium has been easier to access for the study of postviral effects,[11,12,23,24] and the recent increase in biopsy studies of the olfactory epithelium provides us with a detailed picture of the damage. Thus it is tempting (and plausible) to attribute olfactory deficits to epithelial effects alone. However, even though epithelial damage clearly occurs, given the extensive literature on the effects of viruses on the CNS, damage to central olfactory pathways may play a role in the olfactory deficits seen after viral infection.

Necrosis of pseudostratified ciliated columnar epithelium as well as altered ciliary activity in the respiratory epithelium can be produced by influenza viruses[25]; similar processes may be acting on the olfactory neuroepithelium.[26] Olfactory biopsy of one patient with post-URI anosmia indicated extensive scarring and replacement of olfactory epithelium with respiratory epithelium.[23] Other studies of the human olfactory epithelium of patients with viral-induced smell loss have shown a decrease in olfactory receptor cells[11,12]; olfactory cilia, the presumed site of the olfactory transduction process, were frequently absent in the remaining olfactory receptor cells.[11] In both of these studies, the greater the degree of receptor damage that was present, the lower the level of olfactory functioning,[11,12] and the less likely that recovery of olfactory function occurred.[12] If the URI increases the proportion of respiratory epithelium within the olfactory cleft, then even if olfactory receptor cells return to normal through regenerative processes, there may be fewer of them and olfactory ability may only improve to a point (see below).

Several immunohistochemical markers are found in the olfactory epithelium, and two of these, neuron-specific enolase and S-100, a glia-specific marker, may be correlated with improvement in olfactory function. It has been assumed that the epithelial changes in patients with olfactory loss are a consequence of *de*generative processes.[11] Although *re*generation could also account for the epithelial changes and affect immunohistochemical results, at least one study failed to find evidence of cell proliferation, as measured by mouse antiproliferating cell nuclear antigen.[12]

TREATMENT AND PROGNOSIS

A few reports have attempted to reverse the effects of URI on olfactory function. After treatment with topical corticosteroids or oral prednisolone, no improvement was seen in olfactory performance in a group of URI patients, even though three patients showed improved function.[27] These patients were not compared to untreated controls. In another study, oral vitamin B, adenosine triphosphate, and betamethasone nasal drip were used for 1 to 11 months of therapy in 50 patients.[12] While the psychophysical techniques used to evaluate olfactory function were quite different than those used routinely in the United States, these investigators reported recovery in 22% and improvement in another 20%, although again these rates were not compared to untreated controls. This report also described a technique that was reported to distinguish patients likely to improve with therapy. This test determined the presence of an odor sensation in response to an intravenous injection, an interesting technique in that it may selectively evaluate retronasal olfactory ability.[12]

While medical therapies for this common olfactory disorder would be welcomed, there is some consensus that medical regimens known to produce improvement in olfactory function in nasal/sinus disease are ineffective in treating URI-induced anosmia.[2,11] Because medical therapies are unavailable for URI-induced anosmia, many physicians believe that the impairment seen in URI patients in the first year is permanent.[3,11,28] This belief persists despite anecdotal reports of spontaneous recovery.[2,29] A recent long-term follow-up study of patients with post-URI hyposmia or anosmia showed improvement in olfactory function,[9] as measured by self-report and scores on the UPSIT. In follow-up evaluation, 62% of patients reported improved or normal sense of smell. The UPSIT scores also were increased for 90% of patients, and the group mean was significantly higher.[9] The time periods between evaluations in this study were longer than those reported in the literature.[2,3] When some of these patients were tested at 1 year after initial evaluation, there was no improvement. When patients were then retested within 3 years of the first evaluation (which occurred between 1 and 2 years after the onset of the olfactory loss), their UPSIT scores were about 3 points higher for the group mean. For patients whose reevaluation occurred at 4 or 5 years after initial evaluation, UPSIT scores were on average 7 points higher. A significant correlation ($r = +0.552$, $P < 0.01$) was found between the change in UPSIT score and the time between the two UPSIT evaluations.[9]

Thus, the time period at which improvement is assessed is important. In an extensive review of patients reported by Deems et al,[3] no improvement was seen in URI patients, although the time period of the second evaluation is not clear. Mott and Leopold,[2] on the other hand, reported that when URI patients were reevaluated at an average of 26 months after initial evaluation, substantially improved function was seen in 15% of patients. With the even longer follow-up period of the Duncan and Seiden[9] study, improvement was seen in 62% to 90% of patients, depending upon whether subjective or objective methods were used.

As discussed in Duncan and Seiden,[9] the change in olfactory function seen in a group mean may not be the most appropriate metric for evaluating recovery. Yamagishi et al[12] suggested that the patients most likely to improve after URI-induced loss were those with less impairment. To state that recovery from URI-induced anosmia or hyposmia does not occur, based on the average of scores on an olfactory test, denies the variability in olfactory deficits seen in this group of patients. Thus, probability of recovery may be a more useful concept in discussing prognosis with the URI patient.

CONCLUSION

An understanding that recovery can and does occur may elucidate the mechanism of olfactory loss in the URI patient. In biopsy material obtained from post-URI patients, cilia are missing from some of the olfactory receptor cells and there are fewer olfactory receptor cells and nerve bundles.[11,12] In addition, there is a decrease in olfactory epithelium, presumably due to aging,[30] which we know causes a decrease in olfactory function[31] (see chapter 13). Thus, the additional insult of viral invasion to olfactory epithelium already modified by age may explain the higher incidence of olfactory deficits as a result of URI in older individuals; damage to the epithelium caused by a virus, when combined with a mild (and unrecognized) level of deficit due to aging, may create enough deficit to be noticeable to the patient. When there is damage or death of the olfactory receptor cells themselves, regeneration of a new population of cells should occur, with an eventual restoration of function, and yet this regeneration process itself may be compromised by aging, or by other factors that have caused olfactory epithelium to be repopulated by respiratory epithelial cells. All of these issues must be assessed in our estimates of the time course of recovery of olfactory function in the URI patient. As our understanding of the natural history of URI-induced anosmia or hyposmia broadens, we may see the development of appropriate intervention. Research in the past few years[2,9,12] provides an improved prognosis for the URI patient to that which has been proposed before[3,11,12,28] and allows us to provide patients with more accurate information. This will aid in counseling the olfactory-impaired patient.

REFERENCES

1. Henkin RI, Larson AL, Powell RD: Hypogeusia, dysgeusia, hyposmia and dysosmia following influenza-like infection. Ann Otol 84:672–682, 1975.
2. Mott AE, Leopold DA: Disorders in taste and smell. Med Clin North Am 75:1321–1353, 1991.
3. Deems DA, Doty RL, Settle RG, et al: Smell and taste disorders, a study of 750 patients from the University of Pennsylvania Smell and Taste Center. Arch Otolaryngol Head Neck Surg 117:519–528, 1991.
4. Duncan HJ, Smith DV: Clinical disorders of olfaction—a review. In: Doty RL (ed): Handbook of Olfaction and Gustation. New York: Marcel Dekker, 1995.
5. Doty RL, Shaman P, Dann M: Development of the University of Pennsylvania Smell Identification Test: A standardized microencapsulated test of olfactory function. Physio Behav 32:489–502, 1984.
6. Seiden AM, Duncan HJ, Smith DV: Physical diagnosis of taste and smell disorders. Otolaryngol Clin North Am 25:817–835, 1992.
7. Smith DV, Frank RA, Pensak ML, Seiden AM: Characteristics of chemosensory patients and a comparison of olfactory assessment procedures. Chem Senses 12:698, 1987.
8. Donnelly JW, Cain WS, Scott A: Parosmia among patients with olfactory complaints. Chem Senses 14:695, 1989.
9. Duncan HJ, Seiden AM: Long-term follow-up of olfactory loss secondary to head trauma and upper respiratory tract infection. Arch Otolaryngol Head Neck Surg 121:1183–1187, 1995.
10. Duncan HJ, Seiden AM, Paik SI, Smith DV: Differences among patients with smell impairment resulting from head trauma, nasal disease or prior upper respiratory infection. Chem Senses 16:517, 1991.
11. Jafek BW, Hartman D, Eller PM, Johnson EW, Strahan RC, Moran DT: Postviral olfactory function. Am J Rhinol 4:91–100, 1990.
12. Yamagishi M, Fujiwara M, Nakamura H: Olfactory mucosal findings and clinical course in patients with olfactory disorders following upper respiratory viral infection. Rhinology 32:113–118, 1994.
13. Stroop MG: Viruses and the olfactory system. In: Doty RL (ed): Handbook of Olfaction and Gustation. New York: Marcel Dekker, 1995.
14. Mohammed AKH, Magnusson O, Maehlen J, et al: Behavioural deficits and serotonin depletion in adult rats after transient infant nasal viral infection. Neuroscience 35:355–363, 1990.
15. Mohammed AKH, Maehlen J, Magnusson O, Fonnum F, Kristensson K: Persistent changes in behaviour and brain serotonin during ageing in rats subjected to infant nasal virus infection. Neurobiol Aging 13:83–87, 1991.
16. Tomlinson AH, Esiri MM: Herpes simplex encephalitis. J Neurol Sci 60:473–484, 1983.
17. Lycke E, Roos B-E: Studies on increased turnover of brain mono-amines induced by experimental herpes simplex infection. Acta Pathol Microbiol Scand 80:695–701, 1972.
18. Lavi E, Gilden DH, Highkin MK, Weiss SR: The organ tropism of mouse hepatitis virus A59 in mice is dependent on dose and route of inoculation. Lab Anim Sci 36:130–135, 1986.
19. Perlman S, Evans G, Afifi A: Effect of olfactory bulb ablation on spread of a neurotropic coronavirus into the mouse brain. J Exp Med 172:1127–1132, 1990.

20. Winkler WG, Fashinell TR, Leffingwell L, Howard P, Conomy JP: Airborne rabies transmission in a laboratory worker. JAMA 226:1219–1221, 1973.
21. Myers LJ, Nusbaum KE, Swango LJ, Hanrahan LN, Sartin E: Dysfunction of sense of smell caused by canine parainfluenza virus infection in dogs. Am J Vet Res 49:188–190, 1988.
22. Myers LJ, Hanrahan LA, Swango LJ, Nusbaum KE: Anosmia associated with canine distemper. Am J Vet Res 49:1295–1297, 1988.
23. Douek E, Bannister LH, Dotson HC: Recent advances in the pathology of olfaction. Proc Soc Med 68:467–470, 1975.
24. Yamagishi M, Nakano Y: Immunohistochemical studies of olfactory mucosa in patients with olfactory disturbances. Am J Rhinol 3:205–210, 1989.
25. Snow JB Jr: The classification of respiratory viruses and their clinical manifestations. Laryngoscope 79:1485–1493, 1969.
26. Doty RL: A review of olfactory dysfunctions in man. Am J Otolaryngol 1:57–79, 1979.
27. Ikeda K, Sakurada T, Suzaki Y, Takasaka T: Efficacy of systemic corticosteroid treatment for anosmia with nasal and paranasal sinus disease. Rhinology 33:162–165, 1995.
28. Leopold DA, Hornung DE, Youngentob SL: Olfactory loss after upper respiratory infection. In: Getchell TV, Doty RL, Bartoshuk LM, Snow JB Jr (eds): Smell and Taste in Health and Disease. New York, Raven Press, 1991.
29. Goodspeed RB, Catalanotto FA, Gent JF, et al: Clinical characteristics of patients with taste and smell disorders. In: Meiselman HL, Rivlin RS (eds): Clinical Measurement of Taste and Smell. New York, Macmillan, 1986.
30. Paik SI, Lehman MN, Seiden AM, Duncan HJ, Smith DV: Human olfactory biopsy: The influence of age and receptor distribution. Arch Otolaryngol 118:731–738, 1992.
31. Cowart B: Relationships between taste and smell across the adult life span. Ann NY Acad Sci 561:39–55, 1989.

7

Posttraumatic Olfactory Loss

LAURENCE J. DINARDO, M.D., F.A.C.S.
RICHARD M. COSTANZO, Ph.D.

Closed head injury is associated with many severe and often lasting sequelae.[1] Consequently, posttraumatic anosmia is frequently overlooked. As investigation into traumatic olfactory disturbance progresses, both its incidence and importance are increasingly recognized.

Olfactory loss from head trauma was initially described in a number of case reports from the late 19th century.[2-6] Recent work from several chemosensory centers indicates that head trauma accounts for approximately 11% to 19% of all smell and taste disorders.[7-9] These olfactory deficits are important to recognize because alterations in mood, appetite, and olfactory-related tasks can adversely affect the head-injured individual. In particular, the inability to detect gas leaks, smoke, or spoiled foods may present serious personal risk. Knowledge of the mechanisms and consequences of traumatic olfactory loss will improve the physician's ability to diagnose and rehabilitate the head-injured patient.

Although olfactory disturbance can result from minor head trauma, the incidence of posttraumatic anosmia seems to correlate with the severity of head injury. Costanzo and Zasler derived estimates from several studies and found anosmia in 24% to 30% of severely injured patients, 15% to 19% among the moderately injured, and 0% to 16% for individuals with mild head injuries.[10] Partial loss of olfaction, while clinically important, may remain undetected. As a result, recent advances in quantitative olfactory testing have become increasingly important in the diagnosis and treatment of posttraumatic anosmia.[11-13] In addition, data from clinical olfactory function tests are beginning to provide information regarding the epidemiology of olfactory disorders.

ETIOLOGY OF TRAUMATIC OLFACTORY LOSS

Several mechanisms may be responsible for the traumatic loss of olfaction: (1) alterations in the sinonasal architecture preventing odorant access to the intact olfactory sensory epithelium, (2) injury to the olfactory epithelium or axons as a result of tearing and shearing forces, and (3) damage to the olfactory brain parenchyma.

SINONASAL TRAUMA

Trauma to the sinuses and nasal cavity may result in olfactory disturbances. Zusho noted that approximately 11% of posttraumatic olfactory losses resulted from nasal fractures or facial contusions.[14] Unilateral olfactory loss and hyposmia rather than anosmia may occur. Sinonasal trauma as a cause of hyposmia is not common, but it is important because resolution of edema, repair of nasal obstruction, or treatment of sinusitis can result in improved olfaction. The head-injured patient should undergo an otolaryngological evaluation including nasal endoscopy to identify anatomical alterations that prevent odorants from reaching the olfactory cleft (Fig. 7–1). Posttraumatic sinusitis and bony defects relevant to olfactory changes may be further elucidated by computed tomography (CT) studies of the skull base and paranasal sinuses.

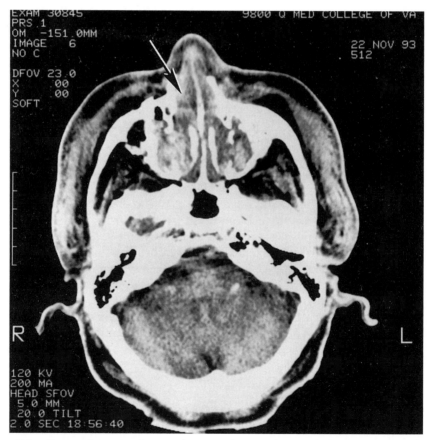

Figure 7–1. Computed tomography scan demonstrating complete nasal airway obstruction (arrow) after head injury with facial trauma.

INJURY TO THE OLFACTORY APPARATUS

Cribriform plate fracture is not a common cause of anosmia. However, the axons of the olfactory receptor cells that pass through the rigid confines of the cribriform plate are susceptible to tearing and shearing forces. The forces most commonly result from rapid shifts in the brain parenchyma secondary to coupe and contracoupe forces. Posterior to anterior directed coupe and contracoupe shearing forces are believed to be the most common cause of posttraumatic olfactory loss.[14] The deficit is usually bilateral and complete. Permanent anosmia is frequently the result.

Although severed axons are able to regenerate, recent endoscopic biopsy techniques by Moran et al have shown that they fail to reach the olfactory bulb, possibly due to scar formation at the cribriform plate.[15] Electron microscopy conducted by these authors on olfactory epithelium from two patients suffering from posttraumatic anosmia demonstrated that the dendrites of olfactory receptors were thickened and shortened and often grew at right angles to their expected paths. Few olfactory vesicles were seen and, when present, lacked olfactory cilia.

The notion that regenerating olfactory axons in patients suffering head trauma are unable to contact and receive the nurturing effects of the olfactory bulb is supported by the work of Chuah and Farbman.[16] Olfactory epithelium grown in culture demonstrated olfactory receptors possessing few olfactory cilia. When a portion of olfactory bulb was added, the olfactory receptors grew axons that made contact with the bulb. Subsequently, the olfactory vesicles demonstrated more cilia than when receptor cells were grown alone. In animal studies where olfactory nerve fibers reestablished contact with the olfactory bulb, olfactory function was restored.[17,18]

In addition to a complete otolaryngological evaluation, patients suspected of suffering anosmia as a result of shearing and tearing of the olfactory filaments should undergo nasal endoscopy. The otolaryngologist should also employ clinical and radiographic techniques to carefully evaluate the entire skull base. Specific abnormalities are sought, such as architectural changes of the cribriform plate, and cerebral spinal fluid (CSF) rhinorrhea. Importantly, CSF rhinorrhea does not necessarily indicate a cribriform plate disruption, as fracture of the adjacent ethmoid roof frequently causes dural tears.[19]

INTRACRANIAL INJURY

The use of olfactory testing in combination with brain-imaging techniques substantiates the concept that parenchymal contusions and brain hemorrhage may be significant causes of posttraumatic olfactory disturbance.[20] Interestingly, damage to the medial frontal and anterior temporal lobes often results in difficulty in odor identification rather than anosmia.[21] If higher-order olfactory neurons in the anterior pyriform cortex, amygdala, and temporal lobe region are injured, associated memory loss and behavioral changes may accompany impairment of olfactory recognition.[21–23] An example of a parenchymal lesion resulting in impaired olfactory

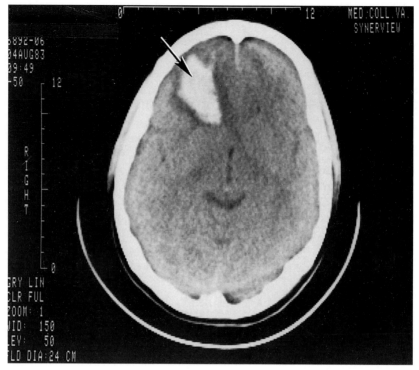

Figure 7–2. Computed tomography scan demonstrating right frontal lobe hematoma (arrow) resulting in impaired olfactory recognition. From Costanzo et al.[10]

recognition is illustrated by the CT scan of a patient with a large frontal lobe hematoma (Fig. 7–2).

Finally, experimental work has demonstrated retrograde degeneration of olfactory neuroepithelium after ablation of the olfactory bulbs.[24] Consequently, injury isolated to the olfactory bulbs is also a plausible central mechanism for posttraumatic anosmia.

COMPREHENSIVE APPROACH TO POSTTRAUMATIC OLFACTORY DISTURBANCES

Given the often insidious nature of posttraumatic olfactory disturbance, it should be apparent that a timely evaluation of the head-injured patient is paramount to successful rehabilitation. The patient history, physical examination, olfactory testing, and appropriate imaging studies are fundamental to the diagnosis.

PATIENT HISTORY

Inquiries should be made into a preexisting olfactory disturbance, prior sinonasal surgery, and antecedent rhinosinusitis. In addition, known nasal polyposis or a recent upper respiratory tract infection must be considered. A history of radiation therapy is important because it has been demonstrated to alter both odor identification and odor detection.[25] Specific questioning focuses on the time course of the olfactory loss. For example, preexisting yet progressive and fluctuating losses are indicative of inflammatory disease or nasal polyposis. Care must be taken, however, not to confuse postnasal drip with a traumatic CSF leak. In conjunction with worsening epistaxis, a sinonasal tumor should be considered.

Once preexisting disease is addressed, a psychosocial history is important to identify medicolegal issues, and also to screen for frontal lobe injury with resultant impairment of both olfaction and cognitive function.[23,26] Associated cranial nerve dysfunction may also exist. Gustatory disturbance is particularly important, as is auditory and vestibular impairment.[14]

Details of the traumatic event should be ascertained. Specifically, the direction of impact and the severity of the head injury provide useful information. External forces applied to the occipital region have most often been associated with anosmia, followed by facial and frontal injuries.[14] The Glascow Coma Scale (GCS) and the duration of posttraumatic amnesia, when used as a measure of the severity of head injury, correlates with the occurrence of olfactory disturbance.[27,28]

Iatrogenic and metabolic causes of olfactory dysfunction must be distinguished from traumatic anosmia. The patient with head trauma is often administered numerous medications. Systemic and topically applied drugs may influence smell function and should be investigated.[29–31] Important categories of drugs known to influence olfaction include certain analgesics, antibiotics, anticholinergics, muscle relaxants, and hypnotics.[26]

Patients suffering head trauma sometimes experience numerous bodily injuries with concomitant multiple organ system failure. Evidence exists that individuals requiring hemodialysis manifest decreased ability to discriminate or recognize odors following therapy.[32]

Head injury may result in epilepsy. Olfactory auras are common events preceding convulsive seizure disorders.[33] Treatment of these disorders has been reported to reduce olfactory hallucinations.[34]

PHYSICAL EXAMINATION

Unfortunately, examination of the patient with head injury does not usually include a sensory evaluation. A complete otolaryngological examination includes motor, sensory, and somatosensory cranial nerve testing. With regard to olfactory testing, the chemosensory function of the trigeminal nerve must be distinguished from that of cranial nerve I. The trigeminal nerve provides extensive innervation to the mouth and nose. Sensory perception is particularly acute in the anterior oral and posterior nasal cavity. Patients may confuse the irritating sensation rendered by the trigeminal nerve with a smell or taste perception.

The rhinologic examination is systematic and complete. External deformities, palpable crepitus, or step-offs are indicative of sinonasal fractures. If olfactory function should improve after the application of decongestant, an inflammatory or obstructive process is sought. Visualization of the inferior, middle, and superior meati are accomplished using the 4.0-mm 30° endoscope. Occasionally the 2.4-mm endoscope is useful in the evaluation of the nasopharynx and olfactory cleft.

OLFACTORY TESTING

Over the past several years a variety of screening and comprehensive olfactory tests have become available. Implementation of olfactory function tests has provided improved identification and, consequently, heightened awareness of posttraumatic olfactory losses.[22,23,35] While posttraumatic olfactory testing is routine at many smell and taste research centers, it remains underutilized in most hospitals.

Olfactory screening tests have been found to correlate with the severity of head injury, as measured by the GCS. Heywood, Zasler, and Costanzo,[23,27] for instance, demonstrated that 13% of patients were anosmic, while 27% exhibited difficulty with odorant identification or detection after mild head injury (GCS 13 to 15). With severe injury 25% of the participants suffered complete loss of olfaction, while 67% demonstrated some dysfunction.

More sophisticated threshold detection testing and odorant identification analyses are useful in further distinguishing and quantifying receptor cell dysfunction and olfactory losses due to cortical injury.[20] Until olfactory testing becomes routine in the evaluation of the head-injured patient the true incidence and impact of posttraumatic olfactory impairment will remain unknown.

IMAGING TECHNIQUES

Plain film radiography is not particularly useful in assessing posttraumatic anosmia. Advances in CT scanning and the development of magnetic resonance imaging (MRI) have, however, proved valuable. High-resolution, thin-cut (3 mm) CT scans in both the axial and coronal planes are useful adjuncts in diagnosing facial bone displacement, posttraumatic sinusitis, and significant cribriform plate disruption. Computed tomography cisternography is particularly helpful in isolating CSF leaks.[36]

The most recent and exciting advances have occurred with MRI. Intracranial injuries long suspected of causing posttraumatic olfactory loss may now be visualized. In particular, frontal lobe contusion and hematoma are delineated by axial and coronal imaging techniques.

TREATMENT AND PROGNOSIS

Nasal obstruction that prevents odorants from reaching an intact olfactory epithelium is one of the few treatable forms of traumatic anosmia. Treatment involves repair of facial fractures, reduction of sinonasal inflammation, and removal of intranasal blood clots. Early recovery from traumatic olfactory loss may result. Iatrogenic and metabolic anosmia occurring during the treatment of the head-injured patient should also be considered reversible. Alterations in drug regimens and the resolution of renal disorders, in particular, may improve olfaction.

Unfortunately, most posttraumatic olfactory disturbances are permanent. Delayed improvement can occur, however, and could result from recovery or regeneration of injured olfactory neurons. Although functional recovery and regeneration of olfactory neurons has been demonstrated in laboratory animals,[17] these processes have not been confirmed in humans. Nevertheless, one third of moderately to severely head-injured patients were found to recover some olfaction in a study by Costanzo and Becker.[37] These authors also note that the likelihood of improvement diminishes with the passage of time. In particular, patients who recover generally do so within 6 months to 1 year after injury.[38]

SOCIAL AND REHABILITATIVE ISSUES

Olfactory loss presents a significant functional impairment that is frequently underestimated. Rigid schedules pertaining to food preparation are important to avoid undetected spoilage. Personal hygiene, homemaking, and child care require close attention. The ample use of smoke detectors is important. Since taste is compromised, menu planning and monitoring of salt intake may be required to ensure appropriate balance of the anosmic individual's diet.

Employment and disability status are significant rehabilitative issues for the head-injured patient.[39,40] Vocational concerns include the reliance on a sense of smell as well as judgment. Frontal lobe injury, in particular, may not only interfere with olfaction but also cause subtle cognitive impairment, an occurrence detrimental to vocations such as a cook or fireman. Varney, in fact, demonstrated a greater than 90% frequency in vocational disability among individuals with traumatic anosmia as opposed to 12% for normosmic head-injured counterparts.[22]

CONCLUSION

Loss of olfaction is frequently overlooked in head-injured patients. Consequently, the true incidence of posttraumatic anosmia is not known. A thorough history and physical examination with regard to olfaction is necessary in all patients suffering head injury and requires an understanding of the pathophysiology of traumatic olfactory loss. Modern imaging technologies and olfactory testing are useful adjuvants when applied properly. The clinician should recognize and treat reversible causes of posttraumatic anosmia. When recovery is unlikely, compensatory strategies, in addition to vocational and safety issues, should be addressed.

REFERENCES

1. Finlayson M, Alan J, Garner SH: Brain injury Rehabilitation: Clinical Considerations. Baltimore: Williams and Wilkins, 1994.
2. Ferrier D: The Functions of the Brain. New York: GP Putnam's Sons, chapter IX, pp 181–211, 1880.
3. Jackson JH: Illustrations of diseases of the nervous system. London Hosp Rep 1:470–471, 1864.
4. Legg JW: A case of anosmia following a blow. Lancet 2:659–660, 1873.

5. Notta A: Recherches sup la perte de l'odorat. Arch Gen Med 15:385–407, 1870.
6. Ogle W: Anosmia or cases illustrating the physiology and pathology of senses of smell. Med Chir Trans 53:263, 1870.
7. Deems DA, Doty RL, Settle RG: Smell and taste disorders, a study of 750 patients from the University of Pennsylvania Smell and Taste Center. Arch Otolaryngol Head Neck Surg 117:519–528, 1991.
8. Leopold DA, Wright HN, Mozell MM: Clinical categorization of olfactory loss. Chem Senses 12:708, 1988.
9. Scott AE: Clinical characteristics of taste and smell disorders. Ear Nose Throat J 68:297, 1989.
10. Costanzo RM, Heywood PG, Ward JD, Young HF: Neurosurgical applications of clinical olfactory assessment. Ann NY Acad Sci 510:242–244, 1986.
11. Cain WS: Testing olfaction in a clinical setting. Ear Nose Throat J 68:322–328, 1989.
12. Cain WS, Gent J, Catalanotto FA, Goodspeed RB: Clinical evaluation of olfaction. Am Med J Otolaryngol 4:252–256, 1983.
13. Doty RL, Shaman P, Dann M: Development of the University of Pennsylvania Smell Identification Test: A standardized microencapsulated test of olfactory function. Physiol Behav 32:489–502, 1984.
14. Zusho, H: Posttraumatic anosmia. Arch Otolaryngol 108:90–92, 1982.
15. Moran DT, Jafek BW, Rowley J III, Eller PM: Electron microscopy of olfactory epithelia in two patients with anosmia. Arch Otolaryngol 111:122–126, 1985.
16. Chuah MI, Farbman AI: Olfactory bulb increases marker protein in olfactory receptors cells. J Neurosci 3:2197–2205, 1983.
17. Costanzo RM: Neural regeneration and functional reconnection following olfactory nerve transection in hamster. Brain Res 361:258–266, 1985.
18. Costanzo RM, Yee K, Koster NL: Functional replacement of sensory neurons and axon projections in the mammalian olfactory system. Soc Neurosci Abstr 19:678, 1993.
19. Raveh J, Vuillemin T, Sutter F: Subcranial management of 395 combined frontobasal-midface fractures. Arch Otolaryngol 114:1114–1122, 1988.
20. Costanzo RM, Zasler ND: Head trauma. In: Getchell TV, Doty RL, Bartoshuk LM, Snow JB Jr (eds): Smell and Taste in Health and Disease. New York: Raven Press, 1991.
21. Levin HS, High WM, Eisenberg HM: Impairment of olfactory recognition after closed head injury. Brain 108:579–591, 1985.
22. Varney NR: Prognostic significance of anosmia in patients with closed-head trauma. J Clin Exp Neuropsychol 10:250–254, 1988.
23. Zasler ND, Costanzo RM, Heywood PG: Neuroimaging correlates of olfactory dysfunction after traumatic brain injury. Abstract Arch Phys Med Rehabil 71:814, 1990.
24. Inamitsu M, Nakashima T, Uemura T: Immunopathology of olfactory mucosa following injury to the olfactory bulb. J Laryngol Otol 104:959, 1990.
25. Ophir D, Guterman A, Gross-Isseroff R: Changes in smell acuity induced by radiation exposure of the olfactory mucosa. Arch Otolaryngol Head Neck Surg 114:853–855, 1988.
26. Mott A, Leopold D: Disorders in taste and smell. Update in Otolaryngology I. Med Clin North Am 75:1321–1353, 1991.
27. Heywood PG, Zasler ND, Costanzo RM: Olfactory screening test for assessment of smell loss following traumatic brian injury. 14th Annual Conference on Rehabilitation of the Brain Injured, Williamsburg, VA, 1990. Abstract.
28. Sumner D: Disturbance of the senses of smell and taste after head injuries. In: Vinken PJ, Bruyn GW (eds): Handbook of Clinical Neurology. Amsterdam: North-Holland Publishing, 1975.
29. Doty RL: A review of olfactory dysfunctions in man. Am J Otolaryngol 1:57–79, 1979.
30. Schiffman SS: Taste and smell in disease (first of two parts). N Engl J Med 308:1275–1280, 1983.
31. Schiffman SS: Taste and smell in disease (second of two parts). N Engl J Med 308:1337–1343, 1983.
32. Conrad P, Corwin J, Katz L, Serby M, LeFavour G, Rotrosen J: Olfaction and hemodialysis: Baseline and acute treatment decrements. Nephron 47:115–118, 1987.
33. Tran-Ba-Huy P, Pialoux P: Les anosmies et les parosmies. Rev Prat 27:2623–2630, 1977.
34. Chitonondh H: Stereotaxic amygdalotomy in the treatment of olfactory seizures and psychiatric disorders with olfactory hallucination. Confin Neurol 27:181–196, 1966.
35. Cain WS, Gent JF, Goodspeed RB, Leonard G: Evaluation of olfactory dysfunction in the Connecticut Chemosensory Clinical Research Center. Laryngoscope 98:83–88, 1988.
36. Colquhoun IA: CT cisternography in the investigation of cerebrospinal fluid rhinorrhea. Clin Radiol 47:403–408, 1993.
37. Costanzo RM, Becker DP: Smell and taste disorders in the head injury and neurosurgery patients. In: Meiselman HL, Rivlin RS (eds): Clinical Measurements of Taste and Smell. New York: MacMillian, 1986.
38. Sumner D: Post-traumatic anosmia. Brain 87:107–120, 1964.

39. American Medical Association: Guides to the Evaluation of Permanent Impairment. Chicago: AMA, 1988.
40. Kraus JF, Fife D, Ramstein K, et al: The relationship of family income to the incidence, external causes, and outcomes of serious brain injury, San Diego County, California. Am J Public Health 76:1345–1347, 1986.

8

Olfactory Loss Secondary to Toxic Exposure

LLOYD HASTINGS, Ph.D.
MARIAN L. MILLER, Ph.D.

Among the many causes leading to decrements in olfactory function, exposure to toxic compounds, especially those airborne, represents a small but important percentage of the cases presenting for evaluation. Generally less than 5% of the cases seen by taste and smell centers are linked to exposure to toxic compounds.[1,2] Although the number of cases is low, they are unique in that frequently these exposures are workplace related and compensation is sought through litigation. Contributing to uncertainty in ascribing a causal relationship in such situations is the paucity of research investigating the effects of exposure to toxic agents on olfactory function. Often the reports available describing the effects of toxic compounds on olfaction are clinical or anecdotal and most of the compounds implicated have not been subjected to rigorous scientific study.

The process of olfaction—the inhalation of air containing volatile, odoriferous compounds that subsequently come into contact with the olfactory receptor neurons in the nasal mucosa—also predisposes the system to contact with any deleterious agent in the inspired air. To counter this vulnerability, the system employs a number of defense mechanisms. The nasal mucosa is capable of secreting antibodies as well as antimicrobial proteins such as lactoferrin and lysozyme to deal with inhaled pathogens[3] (see also chapter 6). The common chemical sense, composed of the fine nerve endings of the trigeminal nerve located throughout the nasal mucosa (as well as the face and cornea), responds to airborne sensory irritants. Upon detection of an irritant, a flight response is initiated by the organism; if escape is not possible, the breathing rate and/or pattern are altered, minimizing the degree of exposure to the airways. The nasal passages are also protected by the presence of phase I and phase II enzymes—cytochrome P-450s, glutathione, and related enzymes—found here in concentrations that can be greater, on a per gram basis, than in the liver or lungs.[4] Since the nasal passages are the first site of internal exposure, these tissues are understandably involved in the detoxification process. However, the resultant metabolites can often be toxic themselves, especially to the olfactory receptor neurons in the olfactory epithelium.[5] That the protective mechanisms in the olfactory epithelium may fail or be circumvented is evidenced by the occurrence of damage to the nasal tissues with subsequent olfactory deficits.

Exposure to many airborne compounds at sufficient concentration results in either annoyance or, at a higher concentration, irritation to the nasal passages[6] (see Fig. 8–1). The concentration that causes annoyance may be so high for some compounds as to not be of practical concern; for others, eg, hydrogen sulfide, even very low concentrations result in sensory irritation. Exposure to levels of irritants sufficient to alter olfactory function usually occurs only after accidental exposure. Under these conditions, *dosis sola facit venenum*[7] (only the dose determines the poison), ie, exposure to most irritants, at high enough concentrations, will result in damage to the olfactory system. Of more widespread concern is the toxicity resulting from chronic exposure to compounds at concentrations that do not elicit the sensory irritant reflex. Under these conditions, damage may be so gradual that changes in function occur slowly and imperceptibly. As a consequence, it is very difficult to establish causality in terms of exposure to specific toxic compounds and resultant olfactory dysfunction, especially when other factors causing dysfunction (viral infections, aging, etc) may occur concurrently.

A number of reviews within the last 10 years have focused on the effects of toxic compounds on olfactory function. Amoore[8] listed more than 120 different compounds and/or industrial processes reported to adversely affect olfactory function (Tables 8–1 and 8–2). Over 75 studies were cited that examined the relationship between exposure and olfactory dysfunction. However, most were case reports or case series, with a majority of the studies published in the 1960s or earlier. Given the lack of good industrial hygiene during this period, reported exposure levels can only be viewed as estimates. Undoubtedly the workers were often exposed to very high concentrations of the compounds, such as occurred with the toxic metal cadmium.

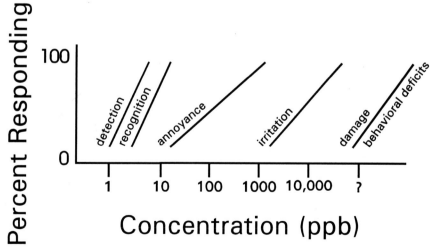

Figure 8–1.   Relationship of odor perception, annoyance, irritation, and olfactory deficits resulting from exposure to increasing concentrations of hydrogen sulfide. From Shusterman[6] and Beauchamp et al.[85]

Cadmium is often cited as the "textbook" example of a toxic compound to which exposure results in anosmia.[9] Examination of the studies[10,11] upon which this claim is based reveals airborne levels were as high as 236 mg Cd/m^3 in certain areas of the plants investigated; the current Occupational Safety and Health Administration (OSHA) Permissible Exposure Limit[12] (PEL—limit on concentration to which a worker may be exposed) for cadmium is 0.005 mg/m^3. This puts into perspective the concentration of cadmium required to produce olfactory dysfunction and the unlikelihood that a worker today would be exposed to comparable levels. Furthermore, many of the referenced studies were deficient in terms of experimental design and the use of appropriate control procedures.

Table 8–1   Substances Affecting Olfaction, for Which There Is Reasonable Supportive Scientific Data

Metallic compounds	Organic compounds
Cadmium compounds	Acetone
Cadmium oxide	Acetophenone
Chromate salts	Benzene
Nickel hydroxide	Benzine
Zinc chromate	Butyl acetate
	Chloromethanes
Metalurgical processes	Ethyl acetate
Chromium plating	Menthol
Lead smelting	Pentachlorophenol
Magnet production	Trichloroethylene
Mercury (chronic intoxication)	
Nickel plating	Dusts
Nickel refining (electrolytic)	Cement
Silver plating	Chemicals
Steel production	Hardwoods
Zinc production	Lime
	Printing
Nonmetallic inorganic compounds	Silicosis
Ammonia	
Carbon disulfide	Manufacturing processes
Carbon monoxide	Acids
Chlorine	Asphalt
Hydrazine	Cutting oils
Nitrogen dioxide	Fragrances
Sulfur dioxide	Lead paints
Fluorides	Paprika
	Spices
	Tobacco
	Varnishes
	Wastewater (refinery)

From Amoore.[8]

Table 8–2   Industrial Substances That May Be Associated With Olfactory Dysfunction in Humans, but for Which There is Little Experimental Data

Metallic compounds
  Alum
  Arsenic compounds
  bis-(Diethyldithiocarbamato)-cadmium
  Chlorovinylarsine chlorides
  Chromic acid
  Copper arsenite
  Dichromates
  Lye
  Osmium tetroxide
  Potassium sulfide
  Silver nitrate
  Strontium sulfate

Metalurgical processes
  Aluminum fumes
  Arsenic
  Chromium fumes
  Copper fumes
  Manganese fumes
  Tin fumes
  Zinc fumes

Nonmetallic inorganic compounds
  Ammonia
  Bromine
  Flue gas
  Hydrazine
  Hydrogen chloride
  Hydrogen fluoride
  Nitric acid
  Nitrogen dioxide
  Phosgene
  Selenium dioxide
  Sewer gas
  Sulfuric acid

Organic compounds
  Acetaldehyde
  Acetic acid
  Acetonitrile
  Acid chlorides
  Benzaldehyde
  Benzoic acid
  Butylene glycol
  Carbon tetrachloride
  Chloroform
  Dimethyl sulfate
  Ethyl ether
  Fluorine compounds
  Formaldehyde
  Fufural
  Halogen compounds
  Iodoform
  Isocyanates
  Nitro compounds
  Phenylene diamine
  Selenium compounds
  Sulfur compounds
  Tetrachloroethane
  Trichloroethane
  War gases (WW I)
  m-Xylene

Manufacturing processes
  Blasting powder
  Coal tar fumes
  Perfumes (concentrated)
  Rubber vulcanization
  Tanning

Dusts
  Cotton
  Cyanides
  Flax
  Flour
  Potash

From Amoore.[8]

A large number of the reports cited by Amoore involved accidental exposure of individuals to compounds at unknown concentrations (Table 8–2). Given that methods of evaluation used in these studies commonly relied on subjective report and that the prior status of olfactory function was not known, the correlation between exposure to these compounds and olfactory deficits is, at best, speculative. Nevertheless, this exhaustive review of the literature contains information that can never be replicated concerning both acute and chronic exposure of humans to high levels of toxic compounds and the subsequent effects on olfactory function.

Cometto-Muñiz and Cain[13] reviewed the effects of exposure to airborne toxicants on upper airway function, focusing to a large extent on the effects of airborne contaminants on the common chemical sense. The reflexive depression in respiratory rate upon exposure to a sensory irritant has been used extensively in the establishment of PELs and Threshold Limit Values (TLVs).* Exposure limits for approximately one third of the compounds regulated by OSHA have been based in part on their irritant properties. Typically, for a new compound with irritant properties, the concentration at which it depresses respiratory rate by 50% is first established (RD_{50}). A value of $0.03 \times RD_{50}$ is then calculated; this value has been found to be very close to the TLV established for many compounds.[13] Whether prolonged exposure at the TLV results in any deficits in olfaction has rarely been determined. Recently, Leopold[14] provided a comprehensive study of the pathophysiology of the nasal cavity resulting from exposure to toxic compounds in the air. Exposure to many of these compounds produces inflammation of the nasal passage, alteration of mucocillary flow, etc, which in turn could have an indirect effect on olfactory function.

More contemporary studies on olfactory function and toxic exposure have been undertaken with greater scientific rigor, but even these are sometimes deficient, given the limitation often inherent in human experimental and/or epidemiological studies. Animal research, in contrast to the human studies, can be more tightly controlled. However, results from animal studies, which often utilize rodents, are difficult to extrapolate to humans. Rodents are macrosmatic animals, while humans are microsmatic; they are also obligate nose breathers, which further limits generalization of the data to humans.

TOXIC COMPOUNDS AND OLFACTION

Although the understanding of olfaction on almost all levels has greatly increased during recent years, damage to the olfactory system due to toxic compounds is still poorly understood. Toxic compounds can affect olfactory function either indirectly, eg, by causing irritation in the nasal passage, blocking the flow of air to the olfactory epithelium, or directly, eg, by altering function or viability of the sensory epithelium. Few compounds have been evaluated specifically for their toxicity to the olfactory

---

* "Threshold limit values refer to airborne concentrations of substances and represent conditions under which it is believed that nearly all workers may be repeatedly exposed day after day without adverse effect." American Conference of Governmental Industrial Hygienists: *1994–1995 Threshold Limit Values for Chemical Substances and Physical Agents and Biological Exposure Indices.* Cincinnati: ACGIH 1994, p. 2.

Table 8–3   The Effects of Toxic Agents on the Olfactory Mucosa of Common Laboratory Rodents (Mice, Rats, Hamsters)

COMPOUND	DURATION	EFFECTS ON HISTOLOGY AND FUNCTION
Acetaldehyde 400–5000 ppm	6 h/d, 5 d/wk 1–28 mo	Degeneration, metaplasia, loss of Bowman's glands and nerve bundles, adenomas, squamous cell carcinomas[56–58]
Acrolein 1.7 ppm	6 h/d 5 d	Hypertrophy, hyperplasia, erosion, ulceration, necrosis, inflammation[59]
Acrylic acid 5–75 ppm	6 h/d, 5 d/wk 13 wk	Degeneration, replacement with respiratory epithelium, inflammation, hyperplasia of Bowman's glands[60]
Benomyl 50–200 mg/m^3	6 h/d, 6 d/wk 13 wk	Degeneration[61]
Bromobenzene 25 µmol/kg IP	[5 min–3 d]	Degeneration of olfactory epithelium and Bowman's glands[62]
Cadmium 250–500 µg/m^3	5 h/d, 5 d/wk 20 wk	Little change[23]
Chlorine gas 0.4–11 ppm	6 h/d, 5 d/wk 16 wk	Degeneration, septal perforations, intracellular deposits of eosinophilic material, mucous cell hypertrophy[63–65]
Chloroform 300 ppm	6 h/d, 7 d	Degeneration of Bowman's glands, cell proliferation in periosteum and bone[66]
Chloropicrin 8 ppm	6 h/d, 5 d	Hypertrophy, hyperplasia, ulceration, necrosis, inflammation[65]
Chlorthiamid 6–50 mg/kg IP	[8 h–7 d]	Degeneration of olfactory epithelium and Bowman's glands, replacement with respiratory epithelium, fibrosis in lamina propria[67]
Dibasic esters 20–900 mg/m^3	4 h/d 7–13 wk	Degeneration, sustentacular cells injured initially, cell proliferation[68,69]
1,2-Dibromo-3-chloropropane 5–60 ppm	6 h/d, 5 d/wk 13 wk	Degeneration, metaplasia, hyperplasia[70]
1,2-Dibromo ethane 3–75 ppm	6 h/d, 5 d/wk 13 wk	Degeneration, metaplasia, hyperplasia[70]
Dichlobenil 12–50 mg/kg IP	[8 h–7 d]	Degeneration of olfactory epithelium and Bowman's glands[5]
1,3-Dichloro propene 30–150 ppm	6 h/d, 5 d/wk 6–24 mo	Degeneration and/or metaplasia[71,72]

Table 8–3   *Continued*

COMPOUND	DURATION	EFFECTS ON HISTOLOGY AND FUNCTION
Dimethylamine 10–511 ppm	6 h/d, 5 d/wk 6–12 mo	Degeneration, loss of nerve bundles, hypertrophy of Bowman's glands[59,65]
1,4-Dithiane 105–420 mg/kg IG	[90 d]	Anisotrophic crystals in giant cells (undetermined chemical composition)[73]
Epichlorohydrin 687 ppm	6 h/d, 5 d	Ulceration, necrosis[65]
Ferrocene 3–30 mg/m³	6 h/d, 5 d/wk 13 wk	Iron accumulation, necrotizing inflammation, metaplasia[74]
Formaldehyde 0.25–15 ppm	6 h/d, 5 d 4 mo	Decreased number of bipolar cells, increased number of basal cells, degeneration of nerve bundles, reduced odor discrimination[34,35]
Furfural 250–400 ppm	7 h/d, 5 d/wk 52 wk	Disorientation of sensory cells, degeneration of Bowman's glands, cystlike structures in lamina propria[75]
Furfural alcohol 2–250 ppm	13 wk	Squamous and respiratory metaplasia of olfactory epithelium, inflammation, hyaline droplets, squamous metaplasia of ducts[76]
Hexamethylene diisocyanate 0.005–0.175 ppm	6 h/d, 5 d/wk 12 mo	Degeneration, mucous hyperplasia[77]
Hydrazine 75–750 ppm	1 h/d 1–10 d	Degeneration[86]
β,β′-Iminodi-propionitrile 200–400 mg/kg IP	[6 h–56 d]	Degeneration of axon bundles, increase of glial fibrillary acidic protein[78]
Methyl bromide 200 ppm	4 h/d, 4 d/wk	Degeneration, decreased carnosine, behavioral deficits[46]
3-Methylfuran 148–322 μmol/I	1 h	Degeneration, more severe in rats than hamsters[79]
3-Methylindole 100–400 mg/kg IP	[7–90 d]	Degeneration, fibrous adhesions, osseous remodeling, Bowman's gland hypertrophy, behavioral deficits[47,80]
Methyl isocyanate 10, 30 ppm	2 h	Degeneration[48]
Napthalene 400–1600 mg/kg IP	[24 h]	Cytotoxicity (mice and hamsters), necrosis (rats)[81]

Table 8–3   *Continued*

COMPOUND	DURATION	EFFECTS ON HISTOLOGY AND FUNCTION
Nickel subsulfide 0.11–1.8 mg/m^3	6 h/d, 5 d/wk 13 wk	Atrophy[82]
Nickel sulfate 3.5–635 mg/m^3	6 h/d, 12–16 consecutive days	Atrophy, degeneration, decrease in carnosine[24]
N-nitroso- dimethylamine 20–80 mg/kg IP	[6 h–30 d]	Degeneration of olfactory epithelium and Bowman's glands[83]
N-nitroso- pyrrolidine 30–100 mg/kg IP	[6 h–30 d]	Degeneration of olfactory epithelium and Bowman's glands[83]
Propylene oxide 10–525 ppm	4 wk	Degeneration[87]
Sulfur dioxide 10–117 ppm	72 h, or 6 h/d, 5 d	Necrosis, edema, destruction, hyperplasia, hypertrophy[27,65]
2,4-Toluene diisocyanate 0.4 ppm	6 h/d, 5 d	Ulceration, necrosis, inflammation, degeneration[65]
3-Trifluoromethyl 1-pyridine 0.1–329 ppm	[6 h–90 d]	Degeneration, reduced Bowman's activity[84]

Durations in brackets are time postexposure until killed.

system, though the effects of toxic compounds on the nasal passages are receiving increased attention. The emphasis usually has been on structural or biochemical/ molecular changes and rarely focused on the behavioral aspects of olfactory function[15,16] (Table 8–3). In addition, there are only a few field/epidemiological studies that have examined the relationship between (occupational) exposure to toxic compounds and olfactory dysfunction. In all, there were less than a dozen studies during the last 20 years that directly addressed this question. The studies described below involved exposure to either toxic metals, irritant gases, or solvents, and in general they employed acceptable experimental designs. The studies cited are not meant to suggest that only these compounds produce toxicity in the olfactory system. The fact is, only a very few compounds have been investigated for deleterious effects on olfactory function.

Toxic Metals

Exposure to cadmium has long been associated with an increased incidence of anosmia in humans. Friberg[17] reported that 37% of workers showed impaired olfactory sense after about 20 years of exposure. Cadmium-exposed workers performed significantly more poorly than age-matched, nonexposed controls on tests of olfactory function.[10] Odor detection thresholds (using serial dilutions of phenol) of 27% of the exposed workers were 200 times greater than normal (ie, "virtual" anosmia), compared to 5% of the controls. Interestingly, about half of these workers with anosmia were unaware of their sensory loss. Physical examination of the nasal mucosa failed to yield any relationship between degree of irritation and olfactory dysfunction (the olfactory epithelium was not examined). Proteinuria was positively correlated with the presence of anosmia, however. Potts[11] found olfactory deficits in approximately 60% of workers exposed to cadmium for 10 to 29 years, and deficits in over 90% of those exposed for more than 30 years. These workers were exposed to extremely high concentrations of cadmium due to poor practices of industrial hygiene, and while conditions like this may not exist now, these reports clearly established that exposure to very high levels of cadmium for long periods of time can result in substantial deficits in olfaction.

More recently, workers exposed to cadmium fumes (from brazing refrigerator coils) were evaluated using both odor detection and odor discrimination tests.[18] Those with the highest body burden of cadmium, determined by measurement of β_2-microgulin in their urine, showed moderate to severe hyposmia, but not anosmia, in the threshold detection task (serial dilutions of *n*-butanol); however, no deficits were observed in the performance of the odor discrimination task. Average duration of exposure to cadmium in these workers was approximately 12 years. The first measurement of airborne cadmium (made 13 years after production began) was 0.300 mg/m³. This is a much lower concentration than that reported in the earlier studies (0.600 to 236.0 mg/m³).[11] The mean duration of worker exposure was also much shorter, but the body burden was still sufficient to alter kidney function. Since the cadmium-related hyposmia was specific only to detection threshold, but not discrimination behavior, the authors concluded that the peripheral olfactory receptor neurons were damaged, and not more central components of the system. In another study,[19] 28% of the workers who had worked 5 years or longer in a cadmium-refining plant reported being anosmic. One limitation of this report was that the classification of anosmia was by self-report; no measures of olfactory function were undertaken in the study. At the time of the study, the average concentration of airborne cadmium was between 0.004 and 0.187 mg/m³; earlier airborne concentrations of cadmium ranged between 0.21 and 0.95 mg/m³. Thus, even moderate exposure to cadmium may adversely affect olfactory function.

Exposure to high levels of organic mercury also produces deficits in olfaction.[20] Patients (mean age = 79 years) suffering from Minamata disease developed through exposure to mercury in utero were tested on both threshold detection tests (serial dilutions of phenyl ethyl alcohol) and smell identification tests (University of Pennsylvania Smell Identification Test [UPSIT]).[21] The UPSIT is a 40-item, standardized, quantitative test of olfactory function that employs microencapsulated odorants in a "scratch and sniff" format. Identification and detection functions decreased with increased age in Minamata disease patients compared to age-matched controls. While no changes were observed in the nasal epithelium in autopsy cases of chronic Minamata disease patients, morphological damage was detected in the olfactory bulb and tract.[22] The exposure levels necessary to produce these effects (and Minimata disease) were reported historically as very high.

Given the difficulty in obtaining tissue samples of human olfactory tissue for analysis, few attempts have been made to examine the olfactory epithelium for damage after exposure to toxic compounds. Alternatively, with an animal model exposure parameters can be precisely controlled, and relevant tissues harvested for examination of histological, biochemical, and molecular indices of insult. Under such conditions, adult male rats were exposed to cadmium (0.500mg/m^3) for 20 weeks but failed to display any evidence of olfactory dysfunction when evaluated on a measure of detection threshold.[23] These levels were sufficient to cause both lung and cardiac toxicity; however, cadmium had little effect on either respiratory or olfactory tissues in the nasal passages. Similar studies have been conducted on nickel and chromium, two additional metals that have been purported to cause olfactory deficits.[8] Although exposure to nickel produced damage in the rodent olfactory epithelium, olfactory function was not affected.[24] There were no apparent consequences of exposure to chromium.[25]

Other studies employing rodents have found numerous industrial compounds that cause damage to the olfactory mucosa (Table 8–3). In general, though, the rodent olfactory system is quite resilient, and in most cases when damage does occur, there is reconstitution of the olfactory epithelium. Although examined less frequently, functional recovery is usually also rapid and complete. While the animal data is important in determining which compounds have the potential for causing damage to the olfactory system of humans, whether recovery proceeds similarly in humans is not known.

Irritant Gases

There is little doubt that accidental exposure to high concentrations of irritant gases can adversely affect olfactory function. Chronic exposure to low levels of airborne contaminants that stimulate the common chemical sense routinely occurs in the workplace, but the effect on olfaction is less clear. Other irritants such as ozone and formaldehyde are common components of both indoor and outdoor pollution.

Japanese workers exposed to SO_2 and/or ammonia (concentrations unknown) reported an increase in subjective complaints involving olfactory disorders.[26] Odor detection thresholds for five different odors as measured with a T & T olfactometer were elevated, but only for the group exposed to SO_2. As with Cd,[18] the deficit appeared to be restricted to detection, and not discrimination behavior. Since both cadmium and SO_2 were airborne contaminants, the site of damage would appear to be peripheral, ie, in the olfactory mucosa. Supporting this assumption is the fact that exposure of mice to SO_2 caused extensive damage in the olfactory epithelium.[27] Many other irritants, including methyl bromide, acrylic acid, aldehydes, and chlorine, produce extensive damage to the olfactory mucosa of rodents *at sufficiently high concentration* (see Table 8–3).

Exposure to ozone, which is a component of both indoor and outdoor pollution, increased olfactory thresholds in human volunteers.[28] After repeated exposure to 0.400 ppb, 4 h/d for 4 days, olfactory thresholds were initially elevated and then returned to normal, suggesting that after the initial insult the system adapted. This situation is quite different from the other studies involving occupational settings, where exposure had occurred for many years. With chronic exposure, it appears that there is an early adaptive phase, which over time with continued exposure can not be maintained, resulting in insult. Thus, duration of exposure and concentration of the toxicant are both important factors in the development of toxicity.

Formaldehyde, an ubiquitous and highly reactive gas, is absorbed mainly in the nose. While exposure to formaldehyde is often alleged to decrease olfactory acuity,[29] the supportive experimental data are very sparse. Workers exposed to 0.075 to 0.750 ppm formaldehyde (alone or in combination with wood dust) showed a slight but significant elevation in detection thresholds (serial dilutions of pyridine) when compared to controls (14.2 vs 15.6).[30] When histology technicians, who were exposed to 0.200 to 1.900 ppm formaldehyde, responded to a questionnaire on symptoms, 68% reported decreased odor perception vs only 9% of the control group.[31] While the questionnaire data are rather imprecise, in conjunction with the threshold study, the overall results support the view that exposure to formaldehyde at relevant domestic and occupational levels can produce deficits in olfactory function.

One area that has received very little attention is the effect of exposure to airborne contaminants on developing organisms. Several studies suggest that exposure of children to low levels of air pollution, including formaldehyde, can adversely affect respiratory function.[32] There is also some evidence in experimental animals suggesting that olfactory function may be affected by formaldehyde. When rodents are exposed to formaldehyde, damage occurs primarily in the anterior of the nasal cavity, sparing the olfactory epithelium.[33] On the other hand, if young rats or ferrets are exposed to low levels of formaldehyde, such exposures result in not only deficits in olfactory function, but also structural changes in the olfactory epithelium.[34,35]

Solvents

Upon inspiration, the nasal mucosa receives maximal exposure to most contaminants in the inhaled air, including volatilized solvents. Due to the lipophilic nature of solvents, after crossing the mucus layer they quickly penetrate the underlying cellular membranes. Many solvents have been found to be neurotoxic,[36] producing effects on the visual, auditory, and olfactory systems. Workers exposed to petroleum products (fuel oil vapors) had higher detection thresholds for *n*-butanol and fuel oil vapor than controls, while thresholds were no different for two other compounds, pyridine and dimethyl disulfide.[37] Although the thresholds of the exposed workers were significantly different than those of the controls, they were still within the normal range. These workers also displayed what the authors termed "odor intensity recruitment," ie, normal perception of strong stimuli but impaired perception of weak stimuli, for all odors tested. The largest decrement was found for detection of fuel oil vapor. Such observations support the concept of *industrial anosmia,* where exposure to strong odors in the workplace results in a reduction in sensitivity for those particular odors, while sensitivity to other odors remains the same. Unlike the deficits resulting from exposure to toxic metals, the effects associated with exposure to solvents appear to be transient.

Probably the most comprehensive investigation addressing the effects of expo-sure to toxic compounds on olfaction was performed by Schwartz et al.[38] Olfactory function of 731 chemical workers exposed to acrylic acid and a variety of acrylates and methacrylates, at levels below the TLVs, was assessed using the UPSIT. While analysis of the cross-sectional data (prevalence) revealed no relationship between chemical exposure and olfactory deficits, a nested case-control study designed to examine cumulative effects of exposure uncovered several associations: (1) olfac-tory dysfunction increased with cumulative exposure, (2) the effects appeared to be reversible, and (3) the highest relative risk of olfactory dysfunction occurred in the group of workers who had never smoked. The more pronounced effects found in the never-smoking group may be due to the fact that smoking induces metabolic en-zymes in the nasal mucosa that provide protection against the toxic effects of exposure to other chemicals.[4] One point stressed by the authors was that although the study suggested exposure to acrylates and methacrylates was associated with a significant decrease in olfaction, the impairment may not always be clinically impor-tant. Whether the observed diminished sense of smell would affect quality of life was not addressed.

In a similar but smaller study, Schwartz et al[39] investigated the effects of solvent exposure on olfactory function in workers at a paint formulation plant. In this study, historic industrial hygiene sampling data were available. Olfactory function was measured using the UPSIT, and the data revealed a subtle, dose-related effect of solvent exposure on olfactory function, but only in never-smokers. The similarity of these results to those found for the chemical workers[38] illustrates how ancillary factors can influence the toxicity of compounds on olfactory function. Although the results were statistically significant, the deficits were very subtle in nature. In spite of the many limitations inherent in field studies, the results suggest that chronic low-level exposure to solvents adversely affects olfactory function. Though olfactory detection thresholds were not determined in these studies, given the greater sensi-tivity of this measure,[18] more robust results might have been obtained if this measure had been used.

Painters may be exposed to higher concentrations of solvents than chemical plant workers. Airborne contaminants are usually closely monitored in commercial plants, as required by government regulations. On the other hand, painters working in confined spaces and without monitoring equipment have the potential to be exposed to high concentrations of solvents. Olfactory function of a group of 54 painters exposed to organic solvents was compared to that of 42 unexposed refer-ents, using the UPSIT.[40] When the influences of age and smoking habits were factored in, no differences in olfactory function were found between the painters and the referents. Odor threshold tests (serial dilutions of pyridine) also revealed no differences between the groups. Lack of effect was attributed to low to moderate degree of exposure to solvents; the small sample size in the painters' study may have further contributed to finding a null effect. Floor-layer workers, characterized as being exposed to high levels of toluene, reported significantly more smell distur-bances than controls when responding by questionnaire.[41]

Levels of airborne contaminants reported in field studies are almost always composite scores developed from available data and best estimates. Furthermore, usually just a subset of the more important confounding variables can be accounted for. The use of volunteers in laboratory studies circumvents many of these limitations, but is itself compromised by restrictions on duration and concentration of exposure, safety and ethical constraints (many compounds are suspected carcinogens), and related factors. Only a single laboratory experiment involving human subjects could be found in the recent literature that explored exposure to a toxic compound and olfactory function.[42] Volunteers were exposed for 7-hour periods to either toluene, xylene, or a mixture according to a Latin square design, and olfactory function evaluated. Detection thresholds for toluene or phenyl-ethyl methyl-ethyl carbinol were measured using the serial dilution technique. A sixfold shift in the detection threshold for toluene was found immediately following exposure to toluene, xylene, or a mixture of the two. No differences were found in phenyl-ethyl methyl-ethyl carbinol thresholds after exposure to any of the compounds. Exposure to a toxic compound may affect detection of the same compound, but have no effect on detection of others. Thresholds for several different compounds should be obtained when testing for olfactory deficits, or at the very least, a test such as the UPSIT should also be employed. In any event, the alterations in olfactory function were reversible, suggesting that the shift in detection threshold was due to olfactory fatigue or adaptation and not to toxic insult to the olfactory receptor neurons. These findings are in agreement with the study by Ahlstrom et al[37] in which workers exposed to fuel oil vapor had the greatest difficulty in detecting fuel oil vapor in threshold tests.

Most reports involving exposure to toxic compounds have used assessment techniques that measure olfactory sensitivity or acuity, eg, detection thresholds. The extent of damage is described as hyposmic, anosmic, etc. While discrimination tasks are also used, eg, UPSIT, performance is described with similar terms. A few reports exist suggesting that exposure to toxic compounds can produce "distortions" in perception, or *dysosmia*. Dysosmia can be classified into *parosmia* (perception of an atypical or distorted odor in response to a particular stimulus) or *phantosmia* (a perceived smell in the absence of an odor) and *aliosmia* (perception of unpleasant odors from nominally pleasant ones).[43] Aliosmias are divided into cacosmia (perception of rotten smell) and torqosmia (perception of burnt or metallic smell). Emmett[44] described the case of a pipefitter exposed to tetrahydrofuran-containing pipe cement who complained of a constant unpleasant smell (parosmia); recovery occurred with removal from exposure to the pipe cement. Shusterman and Sheedy[43] reported that a woman exposed to chloramine gas complained of a "stinging" in the nasopharynx in response to common household odors (torqosmia); recovery occurred without therapy.

MECHANISMS OF TOXIC INSULT/THERAPEUTIC STRATEGIES

The inaccessibility of human olfactory epithelium to biopsy, in conjunction with its small size, has resulted in an almost complete lack of information concerning how toxic compounds affect this specialized region. Measures indirectly related to nasal function, ie, mucocilliary flow, have been utilized, but provide little information on the integrity and operative status of the sense of smell. The development of techniques to collect tissue samples from this area have opened the way for more detailed inspection of the olfactory epithelium[45] (also see chapter 9), but they still have limitations and are used sparingly.

Thus, most of the knowledge concerning mechanisms of toxic action of inhaled compounds is derived from experimental animal studies, which have implicated many agents as olfactotoxins. After exposure to one of these compounds, cytotoxicity is seen almost immediately (>24 hours). Some agents, such as methyl bromide[46] and 3-methyl indole,[47] produce a direct, toxic response. By initially giving low doses of these compounds that induce cytochrome P-450 metabolic enzymes, a protective status is evoked and subsequent, higher exposures produce little or no toxicity. An example of this in humans may be found in the greater susceptibility to solvent-induced olfactory deficits of those chemical workers[37,38] who had never smoked. Other compounds, eg, dimethylamine and 3-methylfuran,[4] require metabolic activation to produce the proximate toxicant. Inhibiting the metabolic enzymes before exposure ameliorates the toxic action of the compound.[5] The exact pathways by which many toxic compounds are metabolized by the nasal mucosa have not yet been elucidated.

After exposure to many olfactotoxins, at least in rodents, the dead cells slough off, and during the next 30 days, the olfactory epithelium is reconstituted. For others, the reconstitution to a mature olfactory epithelium is less rapid, and recuperation of olfactory function takes much longer. One factor that may contribute to this difference is that in the latter situation, the lamina propria and Bowman's glands also appear to be heavily damaged.[48] When there is recovery of the Bowman's glands, function returns.[25] Whether this is due mainly to a return to normal operation by the Bowman's glands or also involves the synthesis and release of necessary growth factors is unknown. What role damage to the Bowman's glands plays in the development of olfactory deficits in humans is also unknown.

Concomitant risk factors may work in a synergistic fashion to enhance the extent of damage the system undergoes. Upper respiratory infections usually produce no long-term adverse effect on olfaction; the same is true with exposure to many toxic agents (at low concentrations). Concurrent exposure to a toxic compound during an upper respiratory infection may result in much greater damage than exposure to either alone, due to weakened defense mechanisms.[27] When patients present with olfactory dysfunction after experiencing either condition, an inquiry should be made as to whether the other was also present.

Attempts to ameliorate the dysfunction resulting from exposure to toxic compounds are limited. Topical steroids, which are routinely used for alleviation of rhinitis and polyposis, may do more than simply restore the airway passage by shrinking inflamed tissue. There is evidence that steroids may hasten the reconstitution of the olfactory epithelium after insult, restoring olfactory function much more quickly than without.[49] Also, the development of olfactory cell cultures and an understanding of the many growth factors involved in initiating and guiding differentiation in these cultures[50] may provide new therapeutic avenues in the future.

MULTIPLE CHEMICAL SENSITIVITY

There is evidence to suggest that the olfactory system plays a role, apart from its traditional sensory function, in mediating or triggering a constellation of symptoms, collectively known as multiple chemical sensitivity (MCS). That is, an aberrant response (MCS) is elicited by the perception of a chemical odorant, as opposed to chemical exposure causing abnormal perception of odors. Although a number of different criteria lists have been developed,[51] there is no single, official description of MCS. In general, with MCS, exposure to many airborne contaminants including solvents, perfumes, cleaning agents, gasoline, pesticides, and paint—almost any "chemical smell"—triggers numerous cross-modality symptoms. These include upper and lower respiratory tract irritation, autonomic arousal (lightheadedness, nausea, anxiety, tachycardia), neurological symptoms (short-term memory loss), and others. The most salient feature is that the exposure levels eliciting these responses are usually far below the TLVs for the compounds and have no effect on the vast majority of people. Often the level of exposure sufficient to induce the symptoms needs to be only at the threshold of odor perception.

In the absence of exposure at levels that produce irritant effects or other traditional toxicological end points, acute odor-related symptoms can occur. An example described by Shusterman[6] involved hydrogen sulfide. Hydrogen sulfide and other sulfur gases, eg, thiopenes, mercaptans, have odor thresholds orders of magnitude lower than levels necessary to elicit classical toxicological or irritant effects. However, these compounds are often associated with annoyance and related symptoms at levels that barely exceed the TLV (see Fig. 8–1). For the individual who displays MCS, the response is greatly exaggerated and can become highly aversive.

In an attempt to explain the elicitation of such acute, odor-triggered symptoms, a number of nontoxicological odor-related mechanisms have been proposed. They have been reviewed by Shusterman[6] and include the following: innate odor aversion; odor-related exacerbation of underlying conditions; odor-related aversive conditioning; and odor-related, stress-induced illness. The last two postulated mechanisms rely heavily on learned associations mediated by classical conditioning processes. Although odor-related aversive conditioning usually, but not always, involves a traumatic exposure episode, the odor-related, stress-induced illness paradigm represents the conditioning of an odor cue with autonomic arousal symptoms resulting from stress. The stress may result from the workplace environment (job pressures, employment insecurity, etc) or from perceived dangers, such as working with "unknown chemicals" or living near a toxic waste site. Boxer has stated, "It has been observed clinically that psychologic reactions to exposure to neurotoxins can be more serious than the direct neurotoxic effects."[52(p425)] In any event, odors are very salient sensory cues, and it is postulated that they quickly become associated with the subjective symptoms. After conditioning, the odor cue alone becomes sufficient to elicit the symptoms. These are working hypotheses, and there are little firm experimental data on the subject.

While the above attempts to explain MCS rely heavily on learned associations, more direct, physiological mechanisms also have been proposed. Meggs[53] has advanced the concept of neurogenic inflammation arising from stimulation of chemical irritant receptors as a possible mechanism to explain the occurrence of chemical sensitivities. Neurogenic inflammation involves the release of mediators, eg, substance P, neurokinen A, and calcitonin gene–related peptide, from sensory nerves to produce vasodilation, edema, and other manifestations of inflammation. The inflammation, which can occur in nerves throughout the body, in turn gives rise to rhinitis, headache, arthritis, and airway disorders. Bell et al[54] have proposed that increased sensitivity to chemical exposure seen in MCS patients results from increased limbic system reactivity resulting from time-dependent sensitization. The limbic system receives direct projections from the olfactory system, and limbic dysfunction is associated with multisystem symptoms involving memory, affect, and autonomic activity. Time-dependent sensitization[55] is hypothesized as the mechanism by which an initial high dose, or repeated low doses, come to induce hypersensitivity to chemical stimuli characteristic of MCS patients. These theories are not necessarily mutually exclusive; some combination of all of these may most accurately describe what is actually occurring.

CONCLUSION

Notwithstanding the large number of reports that suggest exposure to toxic compounds results in olfactory dysfunction, a careful review of the literature, including recent reports, suggests that under normal circumstances exposure to high levels of toxic compounds are required to produce deficits in function. This same conclusion was reached by Schwartz et al, who wrote "studies that do measure exposure quantitatively suggest that hyposmia or anosmia only results from exposure to concentrations above the threshold limit value (TLV)."[38(p617)] That is, exposure to very high levels of a number of toxicants, either due to poor industrial hygiene practices or to accidental exposure, can produce olfactory deficits, but chronic exposure to low levels of the compounds generally does not. While some compounds may produce olfactory fatigue or adaptation[37,42] at sub–TLV exposure levels, these deficits are reversible and are not due to toxic insult. Although this view is presented as an accurate assessment of the empirical data, one caveat must be reiterated, and that is that only a very small number of compounds have actually been tested for their effects on olfaction. Moreover, in very few instances have populations of workers actually been evaluated with respect to olfactory function. Given the fact that many people are unaware of a gradual decline in olfactory function, it is very difficult to link olfactory dysfunction with exposure to toxic compounds. Finally, findings from the animal studies suggest that extensive damage to the olfactory epithelium is required before deficits are manifested (Table 8–3 and Fig. 8–1). Whether the human olfactory epithelium is as resilient is not known.

REFERENCES

1. Deems DA, Doty RL, Settle RG, et al: Smell and taste disorders, a study of 750 patients from the University of Pennsylvania Smell and Taste Center. Arch Otolaryngol Head Neck Surg 117:519–528, 1991.
2. Duncan HJ, Smith DV: Clinical disorders of olfaction. A review. In: Doty RL (ed): Handbook of Olfaction and Gustation. New York: Marcel Dekker, 1994.
3. Mellert TK, Getchell ML, Sparks L, Getchell TV: Characterization of the immune barrier in human olfactory mucosa. Otolaryngol Head Neck Surg 106:181–188, 1992.
4. Reed CJ: Drug metabolism in the nasal cavity: Relevance to toxicology. Drug Metab Rev 25:173–205, 1993.

5. Brandt I, Brittebo EB, Feil VJ, Bakke JE: Irreversible binding and toxicity of the herbicide dichlobenil (2,4-dichlorobenzonitrile) in the olfactory mucosa of mice. Toxicol Appl Pharmacol 103:491–501, 1990.
6. Shusterman D: Critical Review: The health significance of environmental odor pollution. Arch Environ Health 47:76–87, 1992.
7. Deichmann WB, Henschler D, Holmstedt B, Keil G: What is there that is not poison? A study of the *Third Defense* by Paracelsus. Arch Toxicol 58:207–213, 1986.
8. Amoore JE: Effects of chemical exposure on olfaction in humans. In: Barrow CS (ed): Toxicology of the Nasal Passages. Washington, DC: Hemisphere Publishing, 1986.
9. World Health Organization: Principles and Methods for the Assessment of Neurotoxicity Associated With Exposure to Chemicals. Geneva: World Health Organization, 1986.
10. Adams RG, Crabtree N: Anosmia in alkaline battery workers. Br J Industr Med 18:216–221, 1961.
11. Potts CL: Cadmium proteinuria—the health of battery workers exposed to cadmium oxide dust. Ann Occup Hyg 8:55–61, 1965.
12. OSHA (US Department of Labor, Occupational Safety and Health Administration): Occupational Exposure to Cadmium. OSHA 3136. Washington, DC: Occupational Safety and Health Administration, 1992.
13. Cometto-Muñiz JE, Cain WS: Influence of airborne contaminants on olfaction and the common chemical sense. In: Getchell TV, Doty RL, Bartoshuk LM, Snow JB (eds): Smell and Taste in Health and Disease. New York: Raven Press, 1991.
14. Leopold DA: Nasal toxicity: End points of concern in humans. Inhalation Toxicol 6(suppl):23–39, 1994.
15. Barrow CS (ed): Toxicology of the Nasal Passages. Washington, DC: Hemisphere Publishing, 1986.
16. Miller FJ (ed): Nasal Toxicity and Dosimetry of Inhaled Xenobiotics. Washington, DC: Taylor & Francis, 1995.
17. Friberg L: Health hazards in the manufacture of akaline accumulators with special reference to chronic cadmium poisoning. Acta Med Scand 138:1–124, 1950.
18. Rose CS, Heywood PG, Costanzo RM: Olfactory impairment after chronic occupational cadmium exposure. J Occup Med 34:600–605, 1992.
19. Yin-Zeng L, Jin-Xiang H, Cheng-Mo L, Bo-Hong X, Cui-Juan Z: Effects of cadmium on cadmium smelter workers. Scand J Work Environ Health 11:29–32, 1985.
20. Furuta S, Nishimoto K, Egawa M, Ohyama M, Moriyama H: Olfactory dysfunction in patients with Minamata disease. Am J Rhinol 8:259–263, 1994.
21. Doty RL, Shaman P, Dann M: Development of the University of Pennsylvania Smell Identification Test: A standardized microencapsulated test of olfactory function. Physiol Behav 32:489–502, 1984.
22. Mizukoshi K, Watanabe Y, Kato I: Otorhinolaryngological findings in patients with Minamata disease. In: Tsubaki T, Takahashi H (eds): Recent Advances in Minamata Disease Study. Tokyo: Kodansya, 1986.
23. Sun T-J, Miller ML, Hastings L: Effects of inhalation of cadmium on the rat olfactory system: Behavior and morphology. Neurotoxicol Teratol 18:89–98, 1996.
24. Evans JE, Miller ML, Andringa A, Hastings L: Behavioral, histological, and neurochemical effects of nickel (II) on the rat olfactory system. Toxicol Appl Pharmacol 130:209–220, 1995.
25. Hastings L, Andringa A, Miller ML: Exposure of the olfactory system to toxic compounds: Structural and functional consequences. Inhalation Toxicol 6(suppl): 437–440, 1994.
26. Harada N, Fujii M, Dodo H: Olfactory disorder in chemical plant workers exposed to SO_2 and/or NH_3. J Sci Labour 59:17–23, 1983.
27. Giddens WE, Fairchild GA: Effects of sulfur dioxide on the nasal mucosa of mice. Arch Environ Health 25:166–173, 1972.
28. Prah JD, Benignus VA: Decrements in olfactory sensitivity due to ozone exposure. Percept Motor Skills 48:317–318, 1979.
29. Spearman CR: Odors, odorants, and deodorants in aviation. Ann NY Acad Sci 58:40–43, 1954.
30. Holstrom M, Wilhelmsson B: Respiratory symptoms and pathophysiological effects of occupational exposure to formaldehyde and wood dust. Scand J Work Environ Health 14:306–311, 1988.
31. Kilburn KH, Seidman BC, Warshaw R: Neurobehavioral and respiratory symptoms of formaldehyde and xylene exposure in histology technicians. Arch Environ Health 40:229–233, 1985.
32. Soyseth V, Kongerud J, Haarr D, Strand O, Bolle R, Boe J: Relation of exposure to airway irritants in infancy to pervalence of bronchial hyper-responsiveness in schoolchildren. Lancet 345:217–220, 1995.
33. Zwart A, Woutersen RA, Wilmer JWGM, Spit BJ, Feron VJ: Cytotoxic and adaptive effects in rat nasal epithelium after 3-day and 12-week exposure to low concentrations of formaldehyde vapor. Toxicology 51:87–99, 1988.

34. Apfelbach R, Reibenspies M, Schmidt R, Weiler E, Binding N, Camman K: Behavioral and structural modifications in the olfactory epithelium after low level formaldehyde-gas exposure. In: Proceedings of the 3rd International Conference on Role of Formaldehyde in Biological Systems. Methylation and Demethylation Processes. Sopron, Hungary: Hungarian Biochemical Society 1992.

35. Apfelbach R, Weiler E: Sensitivity to odors in Wistar rats is reduced after low-level formaldehyde-gas exposure. Naturwissenschaften 78:221–223, 1991.

36. Abou-Donia MB: Solvents. In: Abou-Donia MB (ed): Neurotoxicology. Boca Raton, FL: CRC Press, 1992.

37. Ahlstrom R, Berglund B, Berglund U, Lindvall T, Wennberg A: Impaired odor perception in tank cleaners. Scand J Work Environ Health 12:574–581, 1986.

38. Schwartz BS, Doty RL, Monroe C, Frye R, Barker S: Olfactory function in chemical workers exposed to acrylate and methacrylate vapors. Am J Public Health 79:613–618, 1989.

39. Schwartz BS, Ford DP, Bolla KI, Agnew J, Rothman N, Bleecker ML: Solvent-associated decrements in olfactory function in paint manufacturing workers. Am J Industr Med 18:697–706, 1990.

40. Sandmark B, Broms I, Lofgren L, Ohlson C-G: Olfactory function in painters exposed to organic solvents. Scand J Work Environ Health 15:60–63, 1989.

41. Hotz P, Tschopp A, Soderstrom D, Holtz J, Boillat M-A, Gutzwiller F: Smell or taste disturbances, neurological symptoms, and hydrocarbon exposure. Int Arch Occup Environ Health 63:525–530, 1992.

42. Mergler D, Beauvais B: Olfactory threshold shift following controlled 7-hour exposure to toluene and/or xylene. Neurotoxicology 13:211–216, 1992.

43. Shusterman DJ, Sheedy JE: Occupational and environmental disorders of the special senses. Occup Med 7:515–542, 1992.

44. Emmett EA: Parosmia and hyposmia induced by solvent exposure. Br J Industr Med 33:196–198, 1976.

45. Lovell MA, Jafek BW, Moran DT, Rowley JC III: Biopsy of human olfactory mucosa: An instrument and a technique. Arch Otolaryngol 108:247–249, 1982.

46. Hastings L, Miller ML, Minnema DJ, Evans J, Radike M: Effects of methyl bromide on the rat olfactory system. Chem Senses 16:43–55, 1991.

47. Turk MAM, Henk WG, Flory W: 3-Methylindole-induced nasal mucosal damage in mice. Vet Pathol 24:400–403, 1987.

48. Uraih LC, Talley FA, Mitsumori K, Gupta BN, Bucher JR, Boorman GA: Ultrastructural changes in the nasal mucosa of Fischer 344 rats and B6C3F1 mice following an acute exposure to methyl isocyanate. Environ Health Perspect 72:77–88, 1987.

49. Kimura Y, Miwa T, Furukawa M, Umeda R: Effects of topical application of steroids on olfactory disturbance in mice. Chem Senses 16;297–302, 1991.

50. Mahanthappa NK, Schwarting GA: Peptide growth factor control of olfactory neurogenesis and neuron survival *in vitro*: Role of EGF and TGF-βs. Neuron 10:293–305, 1993.

51. Neathercott JR, Davidoff LL, Curbow B: Multiple chemical sensitivities syndrome: Toward a working case definition. Arch Environ Health 48:19–26, 1993.

52. Boxer PA: Indoor air quality: A psychosocial approach. J Occup Med 32:425–428, 1990.

53. Meggs MJ: Neurogenic inflammation and sensitivity to environmental chemicals. Environ Health Perspect 101:234–238, 1993.

54. Bell IR, Miller CS, Schwartz GE: An olfactory-limbic model of multiple chemical sensitivity syndrome: Possible relationships to kindling and affective spectrum disorders. Biol Psychiatry 32:218–242, 1992.

55. Antelman SM: Time-dependent sensitization as the cornerstone for a new approach to pharmacotherapy: Drugs as foreign/stressful stimuli. Drug Dev Res 14:1–30, 1988.

56. Appelman LM, Woutersen RA, Feron VJ: Inhalation toxicity of acetaldehyde in rats. I. Acute and subacute studies. Toxicology 23:293–307, 1982.

57. Woutersen RA, Appelman LM, Feron VJ, Van der Heijden CA: Inhalation toxicity of acetaldehyde in rats. II. Carcinogenicity study: Interim results after 15 months. Toxicology 31:123–133, 1984.

58. Woutersen RA, Appelman LM, VanGarderen-Hoetmer A, Feron VJ: Inhalation toxicity of acetaldehyde in rats. III. Carcinogenicity study. Toxicology 41:213–231, 1986.

59. Buckley LA, Morgan KT, Swenbert JA, James RA, Hamm TE, Barrow CS: The toxicity of dimethylamine in F-344 rats and B6C3F1 mice following a 1-year inhalation exposure. Fundam Appl Toxicol 5:341–352, 1985.

60. Miller RR, Ayres JA, Jersey GC, McKenna MJ: Inhalation toxicity of acrylic acid. Fundam Appl Toxicol 1:271–277, 1981.

61. Warheit DB, Kelly DP, Carakostas MC, Singer AW: A 90-day inhalation study with benomyl in rats. Fundam Appl Toxicol 12;333–345, 1989.

62. Brittebo EB, Eriksson C, Brandt I: Activation and toxicity of bromobenzene in nasal tissues in mice. Arch Toxicol 64:54–60, 1990.
63. Wolf DC, Morgan KT, Gross EA, et al: Two-year inhalation exposure of female and male B6C3F1 mice and F344 rats to chlorine gas induces lesions confined to the nose. Fundam Appl Toxicol 24:111–131, 1995.
64. Jiang XZ, Buckley LA, Morgan KT: Pathology of toxic responses to the RD50 concentration of chlorine gas in the nasal passages of rats and mice. Toxicol Appl Pharmacol 71:225–236, 1983.
65. Buckley LA, Jiang XZ, James RA, Morgan KT, Barrow CS: Respiratory tract lesions induced by sensory irritants at the RD50 concentration. Toxicol Appl Pharmacol 74:417–429, 1984.
66. Mery S, Larson JL, Butterworth BE, Wolf DC, Harden R, Morgan KT: Nasal toxicity of chloroform in male f-344 rats and female B6C3F$_1$ mice following a 1-week inhalation exposure. Toxicol Appl Pharmacol 125:214–227, 1994.
67. Brittebo EB, Eriksson C, Feil V, Bakke J, Brandt I: Toxicity of 2,6-dichlorothiobenzamide (Chlorthiamid) and 2,6-dichlorobenzamide in the olfactory nasal mucosa of mice. Fundam Appl Toxicol 17:92–102, 1991.
68. Keenan CM, Kelly DP, Bogdanffy MS: Degeneration and recovery of rat olfactory epithelium following inhalation of dibasic esters. Fundam Appl Toxicol 15:318–393, 1990.
69. Bogdanfy MS, Frame SR: Olfactory mucosal toxicity. Integration of morphological and biochemical data in mechanistic studies: Dibasic ester as an example. Inhalation Toxicol 6(suppl):205–219, 1994.
70. Reznik G, Stinson SF, Ward JM: Respiratory pathology in rats and mice after inhalation of 1,2-dibromo-3-chloropropane or 1,2 dibromoethane for 13 weeks. Arch Toxicol 46:233–240, 1980.
71. Stott WT, Young JT, Calhoun LL, Battjes JE: Subchronic toxicity of inhaled technical grade 1,3-dichloropropene in rats and mice. Fundam Appl Toxicol 11:207–220, 1988.
72. Lomax LG, Stott WT, Johnson KA, Calhoun LL, Yano BL, Quast JF: The chronic toxicity and oncogenicity of inhaled technical-grade 1,3-dichloropropene in rats and mice. Fundam Appl Toxicol 12:418–431, 1989.
73. Schieferstein GJ, Sheldon WG, Cantrell SA, Reddy G: Subchronic toxicity study of 1,4-dithiane in the rat. Fundam Appl Toxicol 11:703–714, 1988.
74. Nikula KJ, Sun JD, Barr EB, et al: Thirteen-week, repeated inhalation exposure of F344/N rats and B6C3F$_1$ mice to ferrocene. Fundam Appl Toxicol 21:127–139, 1993.
75. Feron VJ, Kruysse A: Effects of exposure to furfural vapor in hamsters simultaneously treated with benzo(a)pyrene or diethylnitrosamine. Toxicology 11:127–144, 1978.
76. Miller RA, Mellick PW, Leach CL, Chou BJ, Irwin RD, Roycroft JH: Nasal toxicity in B6C3F1 mice inhaling furfuryl alcohol for 2 or 13 weeks. Toxicologist 11:669A, 1991.
77. Foureman GL, Greenberg MM, Sangha GK, Stuart BP, Shiotsuka RN, Thyssen JH: Evaluation of nasal tract lesions in derivation of the inhalation reference concentration for hexamethylene diisocyanate. Inhalation Toxicol 6(suppl):341–355, 1994.
78. Genter MB, Llorens J, O'Callaghan JP, Peele DB, Morgan KT, Crofton KM: Olfactory toxicity of β,β'-iminodipropionitrile in the rat. J Pharmacol Exp Ther 263:1432–1439, 1992.
79. Morse CC, Boyd MR, Witschi H: The effect of 3-methylfuran inhalation exposure on the rat nasal cavity. Toxicology 30:195–204, 1984.
80. Peele DB, Allison SD, Bolon B, Prah JD, Jensen KF, Morgan KT: Functional deficits produced by 3-methylindole-induced olfactory mucosal damage revealed by a simple olfactory learning task. Toxicol Appl Pharmacol 107:191–202, 1991.
81. Plopper CG, Suverkropp C, Morin D, Nishio S, Buckpitt A: Relationship of cytochrome P-450 activity to Clara cell cytotoxicity. I. Histopathologic comparison of the respiratory tract of mice, rats and hamsters after parenteral administration of naphthalene. J Pharmacol Exp Ther 261:353–363, 1992.
82. Dunnick JK, Elwell MR, Benson JM, et al: Lung toxicity after 13-week inhalation exposure to nickel oxide, nickel subsulfide, or nickel sulfate hexahydrate in F344/N rats and B6C3F$_1$ mice. fundam Appl Toxicol 12:584–594, 1989.
83. Rangga-Tabbu C, Sleight SD: Sequential study in rats of nasal and hepatic lesions induced by N-nitrosodimethylamine and N-nitrosopyrrolidine. Fundam Appl Toxicol 19:147–156, 1992.
84. Gaskell BA, Hext PM, Pigott GH, Hodge MCH, Tinston DJ: Olfactory and hepatic changes following inhalation of 3-trifluoromethyl pyridine in rats. Toxicology 50:57–68, 1988.
85. Beauchamp RO, Bus JS, Popp JA, Boreiko CJ, Andjelkovich DA: A critical review of the literature on hydrogen sulfide toxicity. Crit Rev Toxicol 13:25–97, 1984.
86. Latendresse JR, Marit GB, Vernot EH, Haun CC, Flemming CD: Oncogenic potential of inhaled hydrazine in the nose of rats and hamsters after 1 or 10 1-hr exposures. Toxicol Appl Pharmacol 27:33–48, 1995.
87. Eldridge SR, Bogdanffy MS, Jokinen MP, Andrews LS: Effects of propylene oxide on nasal epithelial cell proliferation in F344 rats. Toxicol Appl Pharmacol 27:25–32, 1995.

9

Olfactory Mucosal Biopsy and Related Histology

BRUCE W. JAFEK, M.D.
EDWARD W. JOHNSON, Ph.D.
PAMELA M. ELLER, M.S.
BRUCE MURROW, M.D., Ph.D.

A detailed analysis of the fine structure of olfactory epithelium is a prerequisite for a thorough understanding of olfactory dysfunction. Accordingly, in 1982, an instrument and technique were developed at the Rocky Mountain Taste and Smell Center (RMTSC) for the safe biopsy of the olfactory epithelium from the intact, living human.[1] The procedure has allowed for a morphologic study of the normal olfactory epithelium, which is necessary in interpretation of changes in the pathological epithelium. This chapter describes the biopsy technique and the histological/immunohistochemical findings as they relate to normal human olfactory epithelium.

The biopsy technique for human olfactory epithelium has been in continuous use over the past 14 years without significant modification, and has been widely disseminated to other centers (at NIH, as well as in Cincinnati, Boston, Baltimore, Pennsylvania, and Japan).[2,3] Over 500 biopsy specimens from approximately 125 patients have been obtained without complication. It should be emphasized that at present this procedure is solely of research value. It should be performed only by otolaryngologists or other clinicians who are specifically versed in the technique and thoroughly familiar with the anatomy of the olfactory cleft. In spite of the absence of serious complications to date, potential risks include hemorrhage, meningitis or encephalitis, cerebrospinal fluid (CSF) leak, or brain injury. Informed consent for this procedure must be obtained.

TOPOGRAPHIC ANATOMY

Man is microsmatic relative to other vertebrates.[4] The human olfactory epithelium constitutes 1.25% of the nasal mucosa, and occupies $2\,cm^2$ of the anterior superior portion of the nasal vault. It overlies the superior nasal septum, the cribriform plate, and the superior aspect of the superior turbinate (Fig. 9–1). There are approximately 6×10^6 total receptors with a population density of 30000 receptors/mm[2].[5] The olfactory epithelium is not uniform, and the total area decreases in a patchy fashion with age.[6,7] The boundary of the epithelium is irregular, with islands of respiratory epithelium interspersed with the olfactory epithelium.[6,8] We therefore take four biopsies, two from each side, assuring a rate of biopsy positive for olfactory epithelium of over 85%.

PRELIMINARY EVALUATION

Before biopsy is obtained, the patient with a chemosensory complaint completes a detailed history and undergoes a focused and pertinent general physical examination.

At the RMTSC this evaluation is initiated by completion of a medical and personal history questionnaire. Since this is a standardized historical review, the data can be entered into a computer database, allowing interinstitutional comparisons and subsequent statistical analyses in reference to the findings observed. The physical examination includes a complete otolaryngological examination of the head and neck region, followed by neurological and general examination as indicated by the results of the first two portions of the evaluation.

The taste and smell function evaluation includes a butyl alcohol threshold test, a 7-item smell identification test,[9] the University of Pennsylvania Smell Identification Test (UPSIT),[10] the Q-tip spatial taste test, a sucrose threshold determination, a citric acid magnitude estimation, and, occasionally, electrogustometry. The details of these testing procedures are not pertinent to this chapter and have been published elsewhere[11] (see chapters 2, 4, and 11).

Radiological evaluation of the olfactory system is limited to coronally oriented computerized tomography (CT) of the ethmoid region. Recent improvements in magnetic resonance imaging (MRI) have allowed visualization of central olfactory pathways.[12,13] Truit and Kelly have reviewed the pertinent neuroembryology and normal anatomy, followed by a detailed consideration of the MRI findings in a wide variety of disorders that may affect the human olfactory system.[13] They conclude that "high-resolution coronal MR images, both before and after contrast material, with and without fat suppression techniques, offer the best opportunity to study the olfactory bulbs and tracts."[13]

Other routine tests (eg, complete blood count, blood chemistries) rarely produce diagnostic results in olfactory dysfunction and are not obtained routinely.[11]

BIOPSY

The first biopsy trials were done in the dissection laboratory on cadavers. This approach is highly recommended for the neophyte. Initial biopsies on humans were obtained under general anesthesia, coincident to nasal surgery (eg, septoplasty, rhinoplasty). A simple procedure under local topical anesthesia evolved as experience accrued, and that technique is used currently.[1]

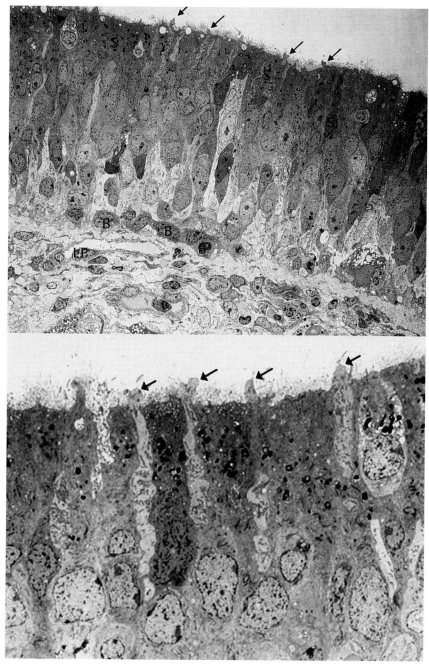

Figure 9–1. Top: Low magnification electron micrograph of olfactory epithelium from a normosmic human volunteer. The epithelium rests on a basement membrane that separates it from the underlying connective tissue of the lamina propria (LP). Basal cells (B) sit on the basement membrane. Vesicles of ciliated olfactory receptor cells (arrows) extend into the nasal cavity. Support cells (S) are interspersed between receptor cells. A microvillar cell (arrowhead) is also present in this field (original magnification ×970). Bottom: Higher-magnification electron micrograph of olfactory epithelium from a normosmic human volunteer. Olfactory vesicles from ciliated olfactory receptor cells (arrows) extend into the nasal cavity. The epithelium is composed primarily of receptor cells with interspersed support cells containing varying amounts of dense pigment (original magnification ×4500).

The donor lies in a supine position. A self-retaining nasal speculum is positioned and the superior nasal cavity is visualized. Initially, fluoroscopy was used to guide the biopsy tool, but this is no longer felt to be necessary. Direct vision with either the operating microscope or endoscope is recommended. The authors prefer the operating microscope, which allows two-handed access, the nondominant hand stabilizing the nasal speculum with or without the endoscope. This technique offers maximal visualization with variable magnification.

The proximity of the biopsy target area to vital structures mandates complete donor cooperation and comfort, along with vasoconstriction of the nasal passage to limit bleeding. Suggestions for anesthesia have included general anesthesia, local topical anesthesia, and no anesthesia. While the latter, as described by Yamagishi et al,[14] is prompted by the concern that the anesthetic might affect the biopsy specimen, this seems unnecessarily conservative and may preclude maximal patient cooperation. We continue to prefer a nasal spray of 4% cocaine hydrochloride for local anesthesia and vasoconstriction. Alternatively, 0.05% oxymetazoline (Afrin) and 0.5% ephedrine can be used, but cocaine, as a simple spray, is thought to be superior. For maximal vasoconstriction and anesthesia at the biopsy site, a cotton pledget soaked in the cocaine solution is inserted superiorly as far as possible into the olfactory cleft.

The olfactory biopsy instrument (OBI) is a hook-shaped open tool that was designed to obtain human olfactory biopsies free of crush artifact. The OBI is commercially available through Storz Instrument Co. (St. Louis). The cocaine pledget is removed and the tool, with its cutting edge facing anteriorly, is advanced superiorly for about 5 to 6 cm until a gentle resistance is encountered at the roof of the olfactory cleft and nasal cavity. The cutting edge is rotated 90° and gently pressed against the mucosa of the septum. With a light downward pull, the biopsy is obtained and the instrument is withdrawn.

To increase the probability of obtaining olfactory epithelium, four biopsies are obtained from each subject, one anterior and one posterior from each side. All biopsies are taken from the relatively flat septum rather than the more convoluted lateral superior turbinate.

Patients are advised not to blow their nose and to avoid heavy lifting or straining for 24 hours after the biopsy. A minimal amount of blood-tinged mucus is to be expected, but the donor is cautioned to return if major bleeding occurs.

Lanza et al[2] have described the use of giraffe forceps (Storz) to obtain olfactory tissue biopsy.[2] They offer illustrative photomicrographs and cite six advantages for the use of the cupped forceps in place of the OBI: (1) "biopsies harvested with the OBI sometimes slipped from the instrument (eg, in cases where they remained tethered to a strand of epithelium)," (2) the use of the giraffe forceps offers "greater control," (3) the giraffe forceps is "widely available," (4) the use of the OBI is "unnecessarily traumatic" because of rotation of the sharp-cutting free edge, (5) the giraffe forceps is "significantly larger" (giving a larger biopsy specimen), and (6) the use of the OBI is "blind." We used a cup forceps initially and had problems with "crush artifact" in the interpretation of histopathological changes. The sharp-cutting OBI solved this problem by cleanly removing a tiny piece of epithelium. The only advantages, in our opinion, of the forceps over the OBI are that the forceps can minimize the occasional loss of the specimen and can obtain a larger specimen.

However, these advantages must be weighed against the trauma to the sample and subsequent interpretive difficulties. Both instruments are easily controlled in the hands of an experienced physician. Both instruments are available commercially, but the OBI is considerably less expensive. The use of the giraffe forceps occasionally resulted in out-fracture of the superior turbinate. This did not occur with the OBI and was likely due to the forceps' greater width (3 mm vs 1.5 mm). It is left to the judgment of the physician which tool causes more trauma for the donor. As Lanza et al[2] noted, "although the olfactory cleft was identified each time a biopsy was taken, the actual biopsy site was *not usually visualized* [emphasis added] during the biopsy procedure. Therefore the techniques employed with both instruments are "blind" to the site of biopsy. The use of nasal forceps[15] and Nakano's forceps (available through Keheller Instrument Company)[16] also has been described. These tools are similar to the giraffe forceps, with comparable advantages and disadvantages. We continue to use the OBI, but recognize that others may have their own preferences and that technology and techniques will continue to evolve.

SPECIMEN PROCESSING

Biopsy specimens are processed for routine electron microscopic evaluation of epithelial fine structure. Each is immediately, placed in a solution of 2% glutaraldehyde and 0.6% paraformaldehyde, buffered to pH 7.2 with 0.06 mol/L sodium cacodylate.[17] Preservation is improved by the addition of 0.05 mol/L $CaCl_2$ to the fixative. Tissues are fixed overnight at room temperature, rinsed in buffer, and embedded in Spurr's low-viscosity resin.[18] Thin sections are cut on an ultramicrotome, mounted on Formvar-coated slot grids,[19] double-stained with uranyl acetate and lead citrate, examined, and photographed with an electron microscope operated at 80 kV.[19]

For the immunocytochemical studies,[20] perfusion or tissue immersion was done using 4% paraformaldehyde and 0.2% picric acid in 0.1 mol/L phosphate buffer (pH 7.3). Tissues were postfixed in the same fixative overnight and transferred to buffer alone. For immunoelectron microscopy, 0.2% glutaraldehyde was added to the original fixative.

After buffer rinses tissues were cryoprotected in a series of glucose and glycerin solutions up to 20% and 15%, respectively. Tissues were frozen onto the stage of a sliding microtome and cut at 50 μm.

MICROSCOPIC STRUCTURE

Nonolfactory Nasal Epithelium

Five types of epithelium line the nasal respiratory passages in humans.

Anteriorly, in direct contact with the environment where it is subject to considerable abrasion and drying, the mucosa is keratinized, stratified squamous epithelium. The submucosal region contains sebaceous glands and sweat glands. Follicles of coarse hairs that aid in filtration also are observed in the submucosa.

Anterolaterally, the sinus cavities are lined by low pseudostratified respiratory epithelium, which is flattened and modified to a simple cuboidal type in some areas and contains a few goblet cells and a very few seromucous glands. The lamina propria is thin and blends in with the underlying periosteum. *Posterolaterally*, the nose loses its filtration and monitoring functions and there is an abrupt transition to a moist, nonkeratinized stratified squamous epithelium. This epithelium is similar to that found throughout the oral cavity.

Superiorly, respiratory epithelium lines the nasal cavity from the posterior nasal vestibule to the nasopharynx. This respiratory epithelium continues up into the olfactory region and interdigitates irregularly with the olfactory epithelium in the roof of the nasal vault.[8] Respiratory epithelium is a pseudostratified columnar epithelium. It consists of ciliated respiratory cells, mucous cells, intermediate cells, and a layer of basal cells that rest on a basement membrane. The epithelium is supported by a lamina propria of loose connective tissue. Within the lamina propria is a unique vascular plexus designed for rapid passage of fluid and dissolved substances between blood vessels and tissues. The lamina propria also contains a collection of lymphocytes that may play a role in the body's defense mechanisms as well as cell-mediated immunological responses. A primary role of respiratory epithelium is the conditioning of air prior to its entry into the lower respiratory passages. This is accomplished by filtration, warming, and humidification.

Olfactory Epithelium

The human olfactory epithelium is a pseudostratified columnar epithelium resting on a highly cellular lamina propria that extends more than 150 μm to the underlying bone and contains Bowman's glands; there is no submucosa. Four major cell types have been identified: ciliated bipolar olfactory receptor cells, microvillar cells, support (sustentacular) cells, and basal cells. With the exception of basal cells, all cell types project to the epithelial surface. Degenerating cells and wandering lymphocytes are present throughout the epithelium.

Olfactory Receptors

Ciliated olfactory receptor cells are bipolar neurons that extend a dendrite toward the epithelial surface and an axon toward the central nervous system. At the epithelial surface, the dendrite forms a knob, the olfactory vesicle, that bears cilia (Fig. 9–2). The cell body tapers into a long axon that travels several centimeters centrally to synapse in the olfactory bulb. The cell measures 40 to 50 μm from the tip of the olfactory vesicle to the base of the cell body.

The olfactory vesicle has 10 to 30 olfactory cilia that project from its surface into the mucous layer that lines the nasal cavity. Basal bodies lie within the lateral and apical margins of the vesicle. Approximately 1.5 μm from the base, the shaft of the cilium tapers to form the distal segment, which is approximately 0.13 μm in diameter and stretches longitudinally into the mucous layer for distances of 100 to 150 μm. Cross-sections of the cilium near its base show a typical 9 + 2 microtubular arrangement of the internal axoneme. Dynein arms are not present, indicating that

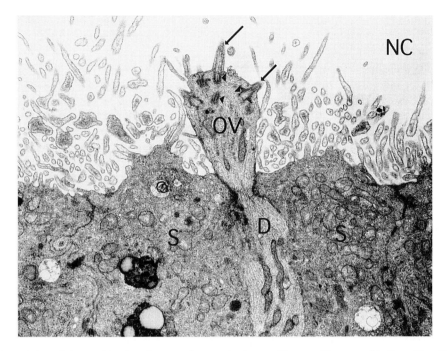

Figure 9–2. High-magnification electron micrograph of normal human olfactory epithelium. The olfactory vesicle (OV) extends from the dendrite (D) of a ciliated olfactory receptor cell into the nasal cavity (NC). Cilia (arrows) on the surface of the vesicle project from basal bodies (arrowheads) within the vesicle. Support cells (S) surround the dendrite, and convoluted microvilli project from their surface into the nasal cavity (original magnification ×15 838).

the cilium is immotile. This is in contrast to the cilia of the respiratory epithelial cells. Distally, the cilium loses its doublet peripheral substructure. Near its tip the cilium has only two or three single microtubules at its core. Occasional cilia have blebs on the distal portions.[21]

Basal bodies lie beneath the epithelial surface and are clustered toward the tip of the dendrite. Toward the cell body the cytoplasm contains numerous microtubules oriented parallel to the cell's long axis. Clusters of mitochondria with longitudinal cristae are found throughout the apical portion of the cell body. Typical smooth endoplasmic reticulum is present in the apical portion of the cell.

Tight junctions and junctional complexes mark the boundaries between olfactory receptor cells and adjacent supporting cells at the epithelial surface. Each dendrite usually is surrounded by the cytoplasm of support cells and is therefore spatially separated from other dendrites. Contact between dendrites has been observed, but adjacent membranous specializations (eg, desmosomes) between dendrites have not been described.

The dendrite joins the cell body 20 to 35 µm beneath the epithelial surface. In the supranuclear region of the cell body, cytoplasmic specializations associated with intense macromolecular synthesis are found. The periphery of the cell body contains cisternae of rough endoplasmic reticulum (rER). At or near the center of this rER lies the forming face of a well-developed Golgi apparatus. Near the Golgi's secretory face are multivesicular bodies and large conspicuous membrane-limited electron-dense vesicles. Bundles of longitudinally oriented microfilaments are located at both the center and periphery of the cell along with occasional mitochondria. The nucleus is uniformly euchromatic.

The basal portion of the cell tapers sharply to form the axon that traverses the basement membrane, turns, and travels along the lamina propria. Adjacent axons coalesce into bundles (fila olfactoria) before passing through the cribriform plate en route to synaptic connections with second-order neurons in the olfactory bulb. The axons are of small caliber within the epithelium, becoming even finer (0.2 to 0.3 μm diameter) in the lamina propria. Larger (eg, 1.25 μm), more peripheral axons contain many mitochondria and microtubules, while smaller (<0.2 μm) more proximal axons contain no mitochondria and only a few longitudinally oriented microtubules. Although all axons appear to be enveloped by the cell membranes of support cells, some axons make direct contact with one another. Synaptic specialization between axons has not been seen.

The cytoplasm of the olfactory neuron is generally lighter and less dense than that of the adjacent support cells. Dispersed free ribosomes are fairly common.

Microvillar Cells

A second cell type that reaches the surface of the epithelium is referred to as the microvillar cell because of its prominent apical microvilli[22,23] (Fig. 9–3). There are approximately 600 000 microvillar cells in the human olfactory epithelium, a number far fewer than that of ciliated olfactory neurons. The ratio of microvillar cells to bipolar neurons is estimated at 1:10. The function of the microvillar cell is unknown. There is debate as to whether or not these represent a second receptor cell type.[23,24]

Although somewhat variable in morphology, microvillar cells generally assume a tadpole shape with a large round nucleus located higher in the epithelium than nuclei of other cell types. The apical membrane of these cells has a tuft of 75 to 100 microvilli that project into the mucous layer lining the nasal cavity. These microvilli are approximately 0.1 μm wide and 1.5 μm long and are frequently branched. There is no internal ultrastructure to the microvilli, nor is there a proximal terminal web, although immediately beneath the apical surface lies an extensive network of fine filaments. Neither cilia nor basal bodies have been seen. The narrow neck of the cell is encircled by a junctional complex, with prominent superficial tight junctions that attach it to adjacent support cell(s) or bipolar neurons.

The microvillar cell is generally pale compared to neighboring cells. This is probably due to an absence of cytoplasmic components. There are few mitochondria; however, the apical cytoplasm contains a well-developed system of sER, occasional cisternae of rER, many free ribosomes, a prominent Golgi apparatus, and occasional clusters of membrane-lined vesicles containing electron-dense material at the periphery that resemble lipofuscin accumulations. Apically, a pair of centrioles can be found. The centrally placed nucleus is almost always located superficially to those of other cell types in the stratification of olfactory epithelial cell nuclei and is pale and euchromatic. A slight peripheral clumping of chromatin is apparent. Nucleoli are not observed.

The base of the cell tapers to a thin, axonlike cytoplasmic process that extends down through the epithelium toward the lamina propria. It has been possible to

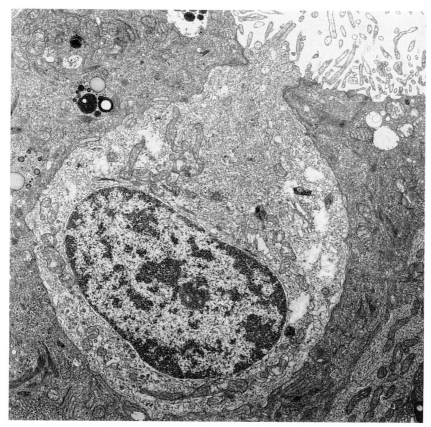

Figure 9–3. High-magnification electron micrograph of normal human olfactory epithelium. A microvillar cell, surrounded by support cells, projects straight short microvilli into the nasal cavity (original magnification ×10690).

trace this process as far as 25 μm, but it has not been observed penetrating the basement membrane or traversing the lamina propria.

Support Cells

The support cells of the olfactory epithelium are tall, columnar cells closely associated with the ciliated bipolar neurons, microvillar cells, basal cells, and each other. At their epithelial surface they are conjoined with their neighboring cells by a typical junctional complex. This consists of a superficial zonula occludens (tight junction), together with a zonula adherens (intermediate junction) and macula adherens (desmosome), producing cell-to-cell adhesion.[25] Support cell spacing appears to be designed to separate adjacent olfactory cells. Junctional complexes may facilitate intercellular metabolic exchange along the sustentacular-olfactory interface. They appear to ensheath dendrite/axon projections deeper in the epithelium. Some dendrites are enveloped by a single support cell, while others are encircled by two or three individual cells.

The surface of support cells contains numerous branched microvilli that emerge from a domeshaped cell apex. No cilia or basal bodies are present. This surface seems quite sensitive to fixation artifacts, often ballooning or seemingly squirting out from the surface of the epithelium. This phenomenon could be part of the contractile olfactomotor response described by Graziadei.[26]

Beneath the surface, support cells exhibit structural polarity, with a dense, organelle-rich cytoplasm at the apex that becomes progressively clear and organelle poor toward the base. The apex of the support cell is packed with organelles including an extensive network of sER, mitochondria, free ribosomes, and membrane-limited vesicular inclusions containing electron-dense, osmophilic material. This either fills the vesicle or is distributed as a halo around the periphery. Occasional vesicles are seen to fuse with the cell surface, although whether they are secreting, pinocytotic, or both is unclear. These vesicles occasionally are characterized as lipofuscin granules, concentrations of olfactory pigment, or lipid-rich granules.[27] Their function, at this point, is unclear, and only a description is provided. Cisternae of rER also are seen, along with a supranuclear Golgi apparatus and occasional lysosomes. The ovoid nucleus, which stains more lightly than nuclei of adjacent bipolar neurons, is euchromatic and more superficially placed than that of the bipolar neuron.

The cytoplasm basal to the nucleus in the lower two thirds of the cell is organelle poor. Large groups of microfilaments organized into wavy straps are present. These large bundles of filaments follow helical pathways along the long axis of the cell. Numerous 500- to 1000-Å vesicles fill the cytoplasm. The base of the supporting cell tapers slightly to rest on the basal lamina.

Functionally, the role of the support cell is conjectural. A supporting role is described. The ultrastructure suggests a phagocytic role or participation in production of the olfactomotor response, or both.

Basal Cells

Basal cells are small polygonal or conical stem cells that differentiate to replace olfactory receptor cells and support cells lost during normal turnover or injury. Moulton's work suggests a turnover rate of 40 days in the mouse.[28] In their resting state, basal cells sit atop the lamina propria and have a central, somewhat darker, heterochromatic ovoid nucleus and a dense cytoplasm with accumulations of filaments. Occasional mitochondria, vesicles (some of which contain osmophilic, electron-dense material), rER and sER, and free ribosomes also are seen.

As they begin to differentiate, their morphology undergoes a profound change. The nucleus becomes more euchromatic, indicating increased cellular transcriptional activity. The basal pole, in the case of bipolar neuron differentiation, gives rise to a slender axon that appears to extend centrally parallel to adjacent axons and axon bundles. The position of the basal cell reflects its chief function, renewal of the bipolar neuron and support cells lost in the course of injury or normal renewal. The superb autoradiographic studies of Graziadei clearly demonstrate this process.[26]

Lamina Propria

The olfactory epithelium is not supported by a true submucosa but lies atop a thin, loose, filamentous basal lamina that lacks the thick collagen and reticular fibers of the basement membrane of the respiratory epithelium.

Deeper than the basal lamina lies a thick, loose network of dense, irregular fibroelastic collagenous connective tissue containing wandering lymphocytes and other inflammatory cells, nerve bundles, and Bowman's glands. Richly vascularized, large venous plexuses are found with less frequency here than in the lamina propria underlying respiratory epithelium. Axons of ciliated bipolar neurons coalesce in the epithelium to enter the lamina propria. They join to form bundles of tiny axons of uniform diameter that pass through the cribriform plate. Most axons are in direct contact with each other, although occasional groups are defined by cytoplasmic invaginations of glial cells that envelope the bundle.

Bowman's Glands

The most conspicuous component of the lamina propria is the tubuloalveolar Bowman's gland. These glands discharge their secretions at the surface of the olfactory epithelium. Viewed in cross-section, the secretory acini are approximately circular and 50 μm in diameter. While some authors have described this as a mixed gland, our observations have identified only serous-producing cells in the human.[28] These cells are pyramidal in shape and surround the lumen. The round to oval nucleus is positioned at the base of the cell, where it is surrounded by abundant cisternae of rER and occasional mitochondria. The base of the cell rests on a basement membrane. Proceeding apically, the supranuclear region contains the Golgi apparatus, while the apical pole is filled with membrane-bound, electron-dense secretory granules. The apical surface of the cell contains short to regular microvilli that are occasionally branched. Other authors describe lysosomes and lipofuscin granules, the source of the yellow pigmentation of the olfactory epithelium.[29] Neither have been apparent in our specimens.

Although the cells of Bowman's glands are clearly exocrine in ultrastructure, designed to secrete protein for export, they do not exhibit the dramatic rER hypertrophy commonly associated with other serous glands, such as the parotid, or the exocrine glands of the pancreas. The nature of the secretion of these glands has not been determined, but because of the location and frequency of occurrence of these glands in olfactory epithelium of other species, the product must be essential to the olfactory transduction process and provides a nonmucous component of the mucous layer in this region.

Undifferentiated basal (stem) cells and slender, filament-packed myoepithelial cells were commonly observed around the periphery of the acini in addition to the serous, secretory cells. The probable function of the myoepithelial cell is contractile. Contractions would squeeze the acinous and thereby move the secretory product along the duct toward the lumen of the nasal cavity.

A more detailed description of the ultrastructure of the olfactory epithelium, along with an extensive comparison to that of the respiratory epithelium and an analysis of the functional significance is available; the foregoing was extensively adapted from the senior author's previous work.[5,22]

IMMUNOCYTOCHEMICAL LABELING CHARACTERISTICS OF THE PERIPHERAL OLFACTORY EPITHELIUM

The olfactory epithelium is a dynamic tissue with the ability to discriminate 10 000 different odors. It also has the unique quality (shared with the vomeronasal neuroepithelium) of replacement of receptor cells as necessary during the life of the organism.

One of the methods used to study the olfactory system has been immunocytochemistry. With this technique, biochemical components of specific cells can be revealed, contributing to understanding how the cell functions.[30] Although possibly all cell types in the olfactory epithelium have been identified with immunocytochemical probes,[31,32] here we will focus on the receptor cells. Some immunocytochemical data that may shed light on the enigmatic olfactory microvillar cell are included.

Immunocytochemistry has demonstrated biochemical similarities among the olfactory receptor neurons of various mammals, including humans. With increasingly sophisticated techniques, the combination of immunocytochemistry with other tools will enhance our understanding of how the olfactory system functions. Following are some of the immunocytochemical traits of the mammalian olfactory epithelium. These were chosen for comparison to the human olfactory system. This section includes reliable markers of olfactory neurons, immunocytochemically identified subsets of these neurons, and molecular markers of presumptive odor receptor and transduction molecules. This section also includes information about the microvillar cess—an egnimatic epithelial cess—and concludes with some work done on human histopathology.

IMMUNOCYTOCHEMICAL "MARKERS" OF OLFACTORY RECEPTOR NEURONS

A prodigious marker of presumably mature olfactory receptor neurons is olfactory marker protein (OMP). An antiserum was generated against OMP two decades ago.[33] Olfactory marker protein immunoreactivity has been useful for identifying differentiated receptor neurons within the mammalian olfactory epithelium,[34,35] including humans.[36,37] Recently, antisera against OMP have also been shown to be effective in identifying the olfactory receptor neurons in other vertebrates,[38,39] indicating the ubiquitous distribution of OMP in olfactory receptor cells of different vertebrate classes.

Through immunocytochemistry, OMP was found in olfactory receptor cells during fetal growth,[40,41] indicating that these neurons mature before birth. Similarly, olfactory neurons in human fetuses express OMP immunoreactivity.[42] As in other mammals, OMP immunoreactivity is first observed in fetal receptor neurons, suggesting the precocious development of these cells.

Complementing the light microscopic observations, immunoelectron microscopy has shown that OMP immunoreactivity extends into the cilia of the receptor neurons.[20,43] These cilia are believed to be the sites of odor transduction. By taking biopsies from patients with olfactory deficits and examining the tissue after exposure to OMP antisera, it may be possible to estimate the number of mature, and presumably functional, receptor neurons. Immunoelectron microscopy may be an aid in determining the morphology of OMP-immunoreactive neurons in various pathologies. To this end, in a preliminary study, we have seen ultrastructural imunolocalization of OMP in olfactory neurons in biopsies taken from normal volunteers (Fig. 9–4).

The mature mammalian olfactory system has the unique ability to generate neurons. Within this dynamic "regenerating" tissue, OMP immunocytochemistry has been used to determine when mature olfactory receptor neurons are replaced.[44,45] A similar turnover of human olfactory receptor neurons has been suggested by a study utilizing OMP immunoreactivity.[46]

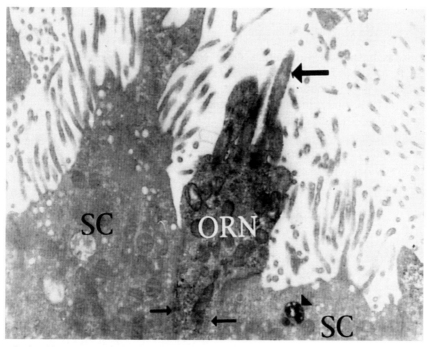

Figure 9–4. Electron micrograph of biopsied olfactory epithelium from a normal human adult volunteer. The tissue was treated with antiserum directed against olfactory marker protein (OMP). OMP immunoreactivity, indicated by electron-dense reaction product, is confined to the receptor neuron (ORN). Reaction product fills the cytoplasm, and is visible in a cilium (large arrow) and dendrite (small arrows). The adjacent support cells (SC) are devoid of label (arrowhead identifies a nonimmunoreactive intrinsic electron-dense pigment within one SC; see Fig. 9–1). This micrograph indicated that immunolocalization of specific proteins can be accomplished with biopsied nasal tissues.

Because human olfactory receptor neurons are OMP immunoreactive, with biopsy, this marker could be used in assessing the status of mature neurons after presumed damage to the tissue. As far as we can tell, OMP immunoreactivity has been used in only one published study of human olfactory pathology.[46]

One caveat concerning OMP is that, although it is used extensively as the standard for the identification of mature olfactory receptor neurons, after biochemical[33] and molecular[47–49] studies on this protein the role of OMP remains unknown.

Another protein found within mammalian olfactory receptor neurons is neuron-specific enolase (NSE). Neuron-specific enolase has been immunocytochemically localized within rodent olfactory neurons.[50] Neuron-specific enolase also has been immunolocalized to olfactory neurons in human fetuses as young as 12 weeks of gestation[51] and in olfactory neurons of the human adult.[50] Recently, NSE immunoreactivity has been used to assess the number of human olfactory receptor neurons following damage to the tissue.[52–54]

A protein that has more recently been discovered as a marker for olfactory receptor neurons is protein gene product 9.5 (PGP). This ubiquitin-associated hydrolase[55] has been detected in mammalian fetal[56,57] and adult[56,58,59] receptor neurons. We have found it to be an excellent probe for rodent (Fig. 9–5) and human fetal and newborn olfactory receptor neurons[(data in prepararation for publication)]. Likewise, PGP also has been used to identify adult human olfactory receptor neurons.[37] The role that PGP plays in the olfactory receptor neurons is unknown.

Among other proteins localized within the olfactory system is carnosine. This protein has been observed within the receptor neurons of humans[60] and other mammals.[61–63] It has been speculated that carnosine may be the neurotransmitter of mammalian olfactory neurons. Recently, immunoelectron microscopy has demonstrated that carnosine is localized to olfactory synaptic terminals in the glomeruli of olfactory bulbs.[64] In that study, glutamate also was immunolocalized to these terminals, while gamma-aminobutyric acid immunoreactivity was detected in intrinsic periglomerular cell dendrites. As the biochemical components in the specific central connections are established, new insights may come to light for novel ways to treat hyposmia or anosmia in patients with idiopathic olfactory deficits.

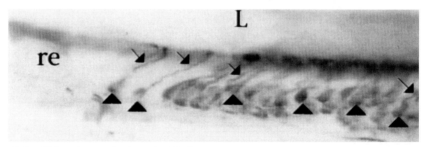

Figure 9–5.   Light micrograph of nasal epithelium taken from a 10 day-old-rat and treated with antiserum against protein gene product 9.5 (PGP). Reaction product is detected only in olfactory receptor cell bodies (arrowheads) and dendrites (arrows), which reach the lumen (L). There are no labeled cells in the adjacent respiratory epithelium (re). This micrograph demonstrates that PGP is localized to olfactory receptor neurons in the nasal epithelium.

IMMUNOCYTOCHEMICALLY IDENTIFIED SUBSETS OF OLFACTORY NEURONS

Although they are structurally homogeneous, there is evidence that olfactory receptor cells may be functionally segregated. Ultimately, this segregation may be related to odor discrimination priorities among the neurons. With a potential association to functional segregation, a number of studies have established that immunocytochemically labeled subsets of olfactory receptor neurons exist within the olfactory epithelium. Subsets of labeled receptor neurons have been shown to project to specific glomeruli in the mammalian olfactory bulb.[65-68] These studies suggest that there is a topographical organization to the connections between olfactory receptor neurons and olfactory bulb glomeruli. In fact, some of these connections may be established during fetal development.[69]

A number of studies have cataloged biochemical moieties in the human olfactory bulb.[36,46,70–76] Similar to other mammals, the human olfactory bulb glomeruli are OMP immunoreactive.[36,46,75] Nevertheless, only one of these studies,[75] based on the observed segregation of OMP-labeled olfactory nerve fascicles, has indicated that human olfactory receptor neurons project to specific glomeruli. Future immunocy-

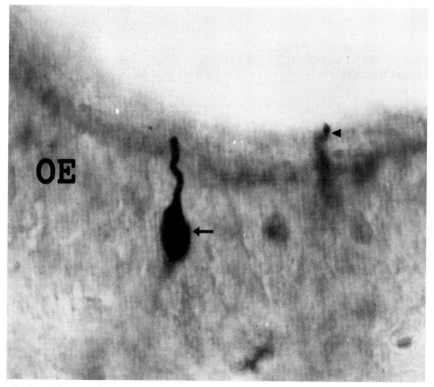

Figure 9–6. Light micrograph of olfactory epithelium (OE) taken from a 2-day-old rat and treated with antiserum against calcium-binding protein D-28k (calbindin). Reaction product is visible throughout the dendrite and cell body (arrow) of one receptor neuron and is seen in the dendritic tip (arrowhead) of a nearby neuron. This micrograph demonstrates that calbindin is expressed in a subset of receptor neurons in the OE of developing rats.

tochemical studies could determine if, in fact, subsets of human olfactory receptor neurons project to specific olfactory bulb glomeruli.

Certain molecules may be primarily important in early development of olfactory receptor neurons. Plank and Mai[77] have shown that a population of olfactory receptor neurons express the 3-fucosyl-*N*-acetyl-lactosamine (CD15) epitope. The number of receptor neurons that express CD15 peaks at 8 days after birth in the rat, and it is found only in a few scattered neurons in the adult. The authors speculated that CD15 may be involved in axon guidance during the proliferation of receptor neurons during development. A somewhat similar pattern, with a different time course, was observed for the intracellular calcium-binding protein D-28k (calbindin) in the rat.[78] The role of calbindin, an intracellular calcium-binding protein, in the olfactory system is unknown, although it may regulate neuronal excitability through calcium sequestering.

We also have observed a population of calbindinlike immunoreactive olfactory receptor neurons in developing rats (Fig. 9–6) and, recently, in developing humans,[(data in prepararation for publication)] with a similar temporal pattern of expression. These findings support the view that there are unique biochemical differences among olfactory receptor neurons early in development and show another developmental similarity between the olfactory receptor neurons of humans and an animal model. It is not yet known if calbindin is expressed in regenerating olfactory neurons.

OLFACTORY ODOR RECEPTORS AND TRANSDUCTION

The recent finding that there is a large gene pool involved with encoding olfactory odor recognition molecules[79] has led to in situ hybridization studies to localize putative olfactory receptor molecules.[80,81] A striking addition to this growing story is the immunocytochemical demonstration that at least one of these odor-recognition proteins may be expressed as early as embryonic day 14 in the rat.[82] This implies that functional segregation of olfactory receptor neurons may be intrinsic to the neurons. Although it is daunting to realize there are likely to be thousands of odor recognition molecules, future studies will have to determine the details of the genesis and neuron population distribution of these molecules to fully understand the olfactory system.

Another one of the challenges in studying the olfactory system is understanding the complexity of the subsequent transduction steps after odorant binding. Progress in identifying components of the olfactory receptor neuron transduction mechanism has included immunocytochemical localization of putative transduction molecules.[83–87]

IMMUNOCYTOCHEMISTRY OF PUTATIVE OLFACTORY MICROVILLAR CELLS

Immunocytochemistry has given us some insight into another cell type in the mammalian olfactory epithelium. This cell, named the microvillar cell (see Fig. 9–3), was identified originally in human adults by our laboratory.[5] The role of the microvillar cell is unknown. Although it has been suggested that the human microvillar cell may contain OMP,[46] we have shown with immunoelectron microscopy that the rat microvillar cell is not immunoreactive for OMP,[20] which conversely suggests that the microvillar cell is not a typical chemoreceptor cell.

On the other hand, recent studies have demonstrated calbindin immunoreactivity within putative microvillar cells in rodents[88–92] and humans[91(our data in prepararation for publication)]. These studies suggest that the microvillar cell may be a neuron or a neuroendocrine cell.

Recently, putative mammalian microvillar cells were shown to be nicotinamide adenine dinucleotide phosphate diaphorase immunoreactive,[93] again linking this cell to neurons.[94] Future immunocytochemical and immunoelectron microscopic studies, in conjunction with other tools, will identify the role of the enigmatic microvillar cell and thus contribute to our understanding of the olfactory system.

PATHOLOGY OF OLFACTORY TISSUE AND IMMUNOCYTOCHEMISTRY

The histopathology of the olfactory epithelium tends to confirm the hypothesis that olfactory dysfunction is accompanied by ultrastructural change that can be correlated with the nature and degree of dysfunction. Herein are discussed immunocytochemical studies that focused on human olfactory pathologies.

Nakashima and colleagues[46] determined that areas of olfactory epithelium from normal adults were devoid of OMP immunoreactivity. These authors suggested that these "histopathological" regions could be where the olfactory epithelium might be replaced by respiratory epithelium. This could ultimately result in hyposmia or anosmia with aging.

Human olfactory epithelium from patients who had Alzheimer's disease shows immunoreactivity for Alzheimer-specific proteins.[95] Autopsied human olfactory bulbs from patients with Alzheimer's disease also show consistent damage.[96] These studies lead to the possibility that olfactory biopsies in conjunction with immunocytochemistry may signal the onset of this disease in individuals with subclinical symptoms.

Another research group has used a number of immunocytochemical markers to compare the labeling characteristics of biopsied olfactory tissue from patients with anosmia due to head trauma,[52–54] choanal atresia, chronic sinusitis, and the common cold.[52] Studies such as these can help in determining the functional status of the olfactory system after various insults.

CONCLUSION

A number of important observations result from a review of the ultrastructure of human olfactory epithelium:

1. Human olfactory epithelium can be successfully biopsied, allowing for detailed ultrastructural studies.

2. Ultrastructural studies of normal human olfactory epithelium have prepared the way for evaluation of the changes that may be occurring in the pathological state.
3. The application of immunocytochemical techniques to normal and pathological olfactory epithelium has important implications for future functional studies.

Acknowledgments—Supported by NIH/NIDCD Grant #DC-00244, The Rocky Mountain Taste and Smell Center.

REFERENCES

1. Lovell MA, Jafek BW, Moran DT, Rowley JC III: Biopsy of human olfactory mucosa: An instrument and a technique. Arch Otol 108:247–249, 1982.
2. Lanza DC, Moran DT, Doty RL, et al: Endoscopic human olfactory biopsy technique: A preliminary report. Laryngoscope 103:815–819, 1993.
3. Yamagishi M, Hasegawa S, Takahashi S, et al: Immunohistochemical method for the diagnosis of olfactory disturbance. Acta Otolaryngol 103:145–150, 1987.
4. Douek E: The Sense of Smell and Its Abnormalities. Edinburgh: Churchill Livingstone, 1974.
5. Moran DT, Rowley JC III, Jafek BW, Lovell MA: The fine structure of the olfactory mucosa in man. J Neurocytol 11:721–746, 1982.
6. Naessen R: An enquiry on the morphological characteristics and possible changes with age in the olfactory region of man. Acta Otolaryngol 71:49–62, 1971.
7. Nakashima T, Kimmelman CP, Snow JB: Structure of human fetal and adult olfactory neuroepithelium. Arch Otol 110:641–646, 1984.
8. Morrison EE, Costanzo RM: Morphology of the human olfactory epithelium. J Comp Neurol 297:1–13, 1990.
9. Cain WS, Gent J, Catalanotto FA, et al: Evaluation of olfactory dysfunction in the Connecticut Chemosensory Clinical Research Center. Laryngoscope 98:83–88, 1988.
10. Doty RL, Shaman P, Dann ML: Development of the University of Pennsylvania Smell Identification Test: A standardized microencapsulated test of olfactory function. Physiol Behav 32:489–502, 1984.
11. Hill DP, Jafek BW: Initial otolaryngologic assessment of patients with taste and smell disorders. ENT J 68:362–370, 1989.
12. Zinreich SJ, Kennedy DW, Rosenbaum AE, et al: CT of nasal cavity and paranasal sinuses: Imaging requirements for functional endoscopic sinus surgery. J Radiol 163:769–775, 1987.
13. Truit CL, Kelly WM: The olfactory system. Neuroimag Clin North Am 3:47–70, 1993.
14. Yamagishi M, Hasegawa S, Hakano Y: Examination and classification of olfactory mucosa in patients with clinical olfactory disturbances. Arch Otol 245:316–320, 1988.
15. Polyzonis BM, Kafandaris PM, Gigis PI, et al: An electron microscopic study of human olfactory mucosa. J Anat 128:77–83, 1979.
16. Trojanowski JQ, Newman PD, Hill WD, et al: Human olfactory epithelium in normal aging, Alzheimer's disease, and other neurodegenerative disorders. J Comp Neurol 310:365–376, 1991.
17. Monti Graziadei GA, Graziadei PPC: Neurogenesis and neuron regeneration in the olfactory system of mammals: II. Degeneration and reconstitution of the olfactory sensory neurons after axotomy. J Neurocytol 8:197–213, 1979.
18. Spurr AR: A low viscosity epoxy resin embedding medium for electron microscopy. J Ultrastruct Res 26:31–43, 1969.
19. Rowley JC III, Moran DT: A simple procedure for mounting wrinkle-free sections on Formvar-coated slot grids. Ultramicroscopy 1:151–155, 1975.
20. Johnson EW, Eller PM, Jafek BW: An immuno-electron microscopic comparison of olfactory marker protein localization in the supranuclear regions of the rat olfactory epithelium and vomeronasal organ neuroepithelium. Acta Otolaryngol 113:766–771, 1993.
21. Frisch D: Ultrastructure of mouse olfactory mucosa. Am J Anat 121:87–120, 1967.
22. Jafek BW: Ultrastructure of human nasal mucosa. Laryngoscope 93:1576–1599, 1983.
23. Rowley JC III, Moran DT, Jafek BW: Peroxidase backfills suggest the mammalian olfactory epithelium contains a second morphologically distinct class of bipolar sensory neuron: The microvillar cell. Brain Res 502:387–400, 1989.

24. Carr VM, Farbman AI, Colletti LM, Morgan JI: Identification of a new non-neuronal cell type in rat olfactory epithelium. Neuroscience 45:433–449, 1991.
25. Farquar MG, Palade GE: Junctional complexes in various epithelia. J Cell Biol 17:375–412, 1963.
26. Graziadei PPC: The ultrastructure of vertebrate olfactroy mucosa. In: Friedman I (ed): The Ultrastructure of Sensory Organs. New York: Elsevier, 1973.
27. Rhodin JAG: Histology. A text and Atlas. New York: Oxford University Press, 1974.
28. Moulton DG, Beidler LM: Structure and function of the peripheral olfactory system. Physiol Rev 47:1–52, 1967.
29. Bloom W, Fawcett DW: A Textbook of Histology. Philadelphia: WB Saunders, 1975.
30. Spicer SS: Advantages of histochemistry for the study of cell biology. Histochem J 25:531–547, 1993.
31. Hempstead JL, Morgan JI: A panel of monoclonal antibodies to the rat olfactory epithelium. J Neurosci 5:438–449, 1985.
32. Hempstead JL, Morgan JI: Monoclonal antibodies reveal novel aspects of the biochemistry and organization of olfactory neurons following unilateral olfactory bulbectomy. J Neurosci 5:2382–2387, 1985.
33. Margolis FL: A brain protein unique to the olfactory bulb. Proc Nat Acad Sci USA 69:1221–1224, 1972.
34. Hartman BK, Margolis FL: Immunofluorescence localization of the olfactory marker protein. Brain Res 96:176–180, 1975.
35. Monti Graziadei GA, Margolis FL, Harding JW, Graziadei PPC: Immunocytochemistry of the olfactory marker proteinm. J Histochem Cytochem 25:1311–1316, 1977.
36. Nakashima T, Kimmelman CP, Snow JB Jr: Olfactory marker protein in the human olfactory pathway. Arch Otolaryngol 111:294–297, 1985.
37. Takami S, Getchell ML, Chen Y, et al: Vomeronasal epithelial cells of the adult human express neuron-specific molecules. NeuroReport 4:375–378, 1993.
38. Rama Krishna NS, Getchell TV, Margolis FL, Getchell ML: Amphibian olfactory receptor neurons express olfactory marker protein. Brain Res 593:295–298, 1992.
39. Riddle DR, Oakley B: Immunocytochemical identification of primary olfactory afferents in rainbow trout. J Comp Neurol 324:575–589, 1992.
40. Farbman AI, Margolis FL: Olfactory marker protein during ontogeny: Immunohistochemical localization. Dev Biol 74:205–215, 1980.
41. Monti Graziadei GA, Stanley RS, Graziadei PPC: The olfactory marker protein in the olfactory system of the mouse during development. Neuroscience 5:1239–1252, 1980.
42. Chuah MI, Zheng DR: Olfactory marker protein is present in olfactory receptor cells of human fetuses. Neuroscience 23:363–370, 1987.
43. Menco BPHM: Electron-microscopic demonstration of olfactory-marker protein with protein G-gold in freeze-substituted, Lowicryl K11M-embedded rat olfactory-receptor cells. Cell Tissue Res 256:275–281, 1989.
44. Harding J, Graziadei PPC, Monti Graziadei GA, Margolis FL: Denervation in the primary olfactory pathway of mice. IV. Biochemical and morphological evidence for neuronal replacement following nerve section. Brain Res 132:11–28, 1977.
45. Samanen DW, Forbes WB: Replication and differentiation of olfactaory receptor neurons following axotomy in the adult hamster: A morphometric analysis of postnatal neurogenesis. J Comp Neurol 225:201–211, 1984.
46. Nakashima T, Kimmelman CP, Snow JB Jr: Immunohistopathology of human olfactory epithelium, nerve and bulb. Laryngoscope 95:391–396, 1985.
47. Margolis FL: Olfactory marker protein: From PAGE band to cDNA clone. Trends Neurosci 8:542–546, 1985.
48. Rogers KE, Dasgupta P, Gubler U, Grillo M, Khew-Goodall YS, Margolis FL: Molecular cloning and sequencing of a cDNA for olfactory marker protein. Proc Natl Acad Sci USA 84:1704–1708, 1987.
49. Sydor W, Teitelbaum Z, Blacher R, Sun S, Benz W, Margolis FL: Amino acid sequence of a unique neuronal protein: Rat olfactory marker protein. Arch Biochem Biophys 249:351–362, 1986.
50. Yamagishi M, Hasegawa S, Takahashi S, Nakano Y, Iwanaga T: Immunohistochemical analysis of the olfactory mucosa by use of antibodies to brain proteins and cytokeratin. Ann Otol Rhinol Laryngol 98:384–388, 1989.
51. Takahashi S, Iwanaga T, Takahashi Y, Nakano Y, Fujita T: Neuron-specific enolase, neurofilament protein and S-100 protein in the olfactory mucosa of human fetuses. An immunohistochemical study. Cell Tissue Res 238:231–234, 1984.
52. Yamagishi M, Hasegawa S, Takahashi S, Nakano Y, Iwanaga T: Immunohistochemical method for the diagnosis of olfactory disturbance. Acta Otolaryngol 103:145–150, 1987.

53. Yamagishi M, Nakamura H, Suzuki S, Hasegawa S, Nakano Y: Immunohistochemical examination of olfactory mucosa in patients with olfactory disturbance. Ann Otol Rhinol Laryngol 99:205–210, 1990.

54. Yamagishi M, Okazoe R, Ishizuka Y: Olfactory mucosa of patients with olfactory disturbance following head trauma. Ann Otol Rhinol Laryngol 103:279–284, 1994.

55. Wilkinson KD, Lee K, Deshpande S, Duerksen-Hughes P, Boss JM, Pohl J: The neuron-specific protein PGP 9.5 is a ubiquitin carboxyl-terminal hydrolase. Science 246:670–673, 1989.

56. Johnson EW, Eller PM, Jafek BW: Protein gene product 9.5 in the developing and mature rat peripheral olfactory and vomeronasal systems. 15th annual meeting of the Association for Chemoreception Sciences, 1993, Abstract 209.

57. Kent C, Clarke P: The immunolocalisation of the neuroendocrine specific PGP 9.5 during neurogenesis in the rat. Dev Brain Res 58:147–150, 1991.

58. Iwanaga T, Han H, Kanazawa H, Fujita T: Immunohistochemical localization of protein gene product 9.5 (PGP 9.5) in sensory paraneurons of the rat. Biomed Res 13:225–230, 1992.

59. Taniguchi K, Saito H, Okamura M, Ogawa K: Immunohistochemical demonstration of protein gene product 9.5 (PGP 9.5) in the primary olfactory system of the rat. Neurosci Lett 156:24–26, 1993.

60. Sakai M, Ashihara M, Nishimura T, Nagatsu I: Carnosine-like immunoreactivity in human olfactory mucosa. Acta Otolaryngol 109:450–453, 1990.

61. Biffo S, Grillo M, Margolis FL: Cellular localization of carnosine-like and anserine-like immunoreactivities in rodent and avian central nervous system. Neuroscience 35:637–651, 1990.

62. Ferriero D, Margolis FL: Denervation in the primary olfactory pathway of mice. II. Effects on carnosine and other amine compounds. Brain Res 94:75–86, 1975.

63. Sakai M, Yoshida M, Karasawa N, Teramura M, Ueda H, Nagatsu I: Carnosine-like immunoreactivity in the primary olfactory neuron of the rat. Experientia 43:298–300, 1987.

64. Sassoè-Pognetto M, Cantino D, Panzanelli P, et al: Presynaptic co-localization of carnosine and glutamate in olfactory neurones. NeuroReport 5:7–10, 1993.

65. Fujita S, Mori K, Imamura K, Obata K: Subclasses of olfactory receptor cells and their segregated central projections demonstrated by a monoclonal antibody. Brain Res 326:192–196, 1985.

66. Key B, Akeson RA: Distinct subsets of sensory olfactory neurons in mouse: Possible role in the formation of the mosaic olfactory projection. J Comp Neurol 335:355–368, 1993.

67. Mori K, Fujita SC, Imamura K, Obata K: Immunohistochemical study of subclasses of olfactory nerve fibers and their projections to the olfactory bulb in the rabbit. J Comp Neurol 242:214–229, 1985.

68. Schwarting GA, Crandall JE: Subsets of olfactory and vomeronasal sensory epithelial cells and axons revealed by monoclonal antibodies to carbohydrate antigens. Brain Res 547:239–248, 1991.

69. Schwarting GA, Deutsch G, Gattey DM, Crandall JE: Glycoconjugates are stage- and position-specific cell surface molecules in the developing olfactory system, 1: The CC1 immunoreactive glycolipid defines a rostrocaudal gradient in the rat vomeronasal system. J Neurobiol 23:120–129, 1992.

70. Nagao M, Oka N, Kamo H, Akiguchi I, Kimura J: *Ulex europaeus I* and glyucine max bind to the human olfacatory bulb. Neurosci Lett 164:221–224, 1993.

71. Ohm TG, Braak E, Probst A: Somatostatin-14-like immunoreactive neurons and fibres in the human olfactory bulb. Anat Embryol 179:165–172, 1988.

72. Ohm TG, Braak E, Probst A, Weindl A: Neuropeptide Y-like immunoreactive neurons in the human olfactory bulb. Brain Res 451:295–300, 1988.

73. Ohm TG, Müller NU, Braak E: Glutamic-acid-decarboxylase- and parvalbumin-like-immunoreactive structures in the olfactory bulb of the human adult. J Comp Neurol 291:1–8, 1990.

74. Ohm TG, Müller NU, Braak E: Calbindin-D-28k-like immunoreactive structures in the olfactory bulb and anterior olfactory uncleus of the human adult: Distribution and cell typology—partial complementarity with parvalbumin. Neuroscience 42:823–840, 1991.

75. Smith RL, Baker H, Kolstad K, Spencer DD, Greer CA: Localization of tyrosine hydroxylase and olfactory marker protein immunoreactivities in the human and macaque olfactory bulb. Brain Res 548:140–148, 1991.

76. Smith RL, Baker H, Greer CA: Immunohistochemical analyses of the human olfactory bulb. J Comp Neurol 333:519–530, 1993.

77. Plank J, Mai JK: Developmental expression of the 3-fucosyl-N-acetyl-lactosamine/CD15 epitope by an olfactory receptor cell subpopulation and in the olfactory bulb of the rat. Dev Brain Res 66:257–261, 1992.

78. Abe H, Watanabe M, Kondo H: Developmental changes in expression of a calcium-binding protein (Spot 35-calbindin) in the nervus terminalis and the vomeronasal and olfactory receptor cells. Acta Otolaryngol 112:862–871, 1992.

79. Buck L, Axel R: A novel multigene family may encode odorant receptors: A molecular basis for odor recognition. Cell 65:175–187, 1991.

80. Ressler KJ, Sullivan SL, Buck LB: A zonal organization of odorant receptor gene expression in the olfactory epithelium. Cell 73:597–609, 1993.
81. Vassar R, Ngai J, Axel R: Spatial segregation of odorant receptor expression in the mammalian olfactory epithelium. Cell 74:309–318, 1993.
82. Koshimoto H, Katoh K, Yoshihara Y, Nemoto Y, Mori K: Immunohistochemical demonstration of embryonic expression of an odor receptor protein and its zonal distribution in the rat olfactory epithelium. Neurosci Lett 169:73–76, 1994.
83. Anholt RRH, Mumby SM, Stoffers DA, Girard PR, Kuo JF, Snyder SH: Transduction proteins of olfactory receptor cells: Identification of guanine nucleotide binding proteins and protein kinase C. Biochemistry 26:788–795, 1987.
84. Asanuma N, Nomura H: Cytochemical localization of adenylate cyclase activity in rat olfactory cells. Histochem J 23:83–90, 1991.
85. Asanuma N, Nomura H: Cytochemical localization of cyclic 3′,5′-nucleotide phosphodiesterase activity in the rat olfactory mucosa. Histochem J 25:348–356, 1993.
86. Menco BPM, Bruch RC, Dau B, Danho W: Ultrastructural localization of olfacatory transduction components: The G protein subunit G_{olf_∂} and type III adenylyl cyclase. Neuron 8:441–453, 1992.
87. Shinohara H, Kato K, Asano T: Differential localization of G proteins, G_i and G_o, in the olfactory epithelium and the main olfactory bulb of the rat. Acat Anat 144:167–171, 1992.
88. Iwanaga T, Takahashi-Iwanaga H, Fujita T, Yamakuni T, Takahashi Y: Immunohistochemical demonstration of a cerebellar protein (Spot 35) in some sensory cells of guinea pigs. Biomed Res 6:329–334, 1985.
89. Johnson EW, Eller PM, Jafek BW: Calbindin-like immunoreactivity in the receptor cells of the rat vomeronasal organ and in isolated cells of the olfactory epithelium. 13th annual meeting of the Association for Chemoreception Sciences, 1991, Abstract 271.
90. Yamagishi M, Nakamura H, Takahashi S, Nakano Y, Kuwano R: Olfactory receptor cells: Immunocytochemisrty for nervous system-specific proteins and re-evaluation of their precursor cells. Arch Histol Cytol 52(suppl):375–381, 1989.
91. Yamagishi M, Nakamura H, Igarashi S, Nakano Y, Kuwano R: Immunohistochemical study of the fourth cell type in the olfactory epithelium in guinea pigs and in a patient. J Otorhinolaryngol Relat Spec 54:85–90, 1992.
92. Yamagishi M, Ishizuka Y, Fujiwara M, et al: Distribution of calcium binding proteins in sensory organs of the ear, nose and throat. Acta Otolaryngol Suppl 506:85–89, 1993.
93. Dellacorte C, Kalinoski DL, Huque T, Wysocki L, Restrepo D: Localization of nitric oxide synthase in the olfactory epithelium of the rat and channel catfish. 16th annual meeting of the Association for Chemoreception Sciences, 1994, Abstract 72.
94. Bredt DS, Snyder SH: Nitric oxide, a novel neuronal messenger. Neuron 8:3–11, 1992.
95. Talamo BR, Rudel RA, Kosik KS, et al: Pathological changes in olfactory neurons in patients with Alzheimer's disease. Nature 337:736–739, 1989.
96. Ohm TG, Braak H: Olfactory bulb changes in Alzheimer's disease. Acta Neuropathol 73:365–369, 1987.

10

Basic Anatomy and Physiology of Taste

DAVID V. SMITH, Ph.D.

The term *taste* is used commonly to refer to the complex of sensations known as *flavor* perception, which includes sensations from the olfactory, gustatory, and trigeminal sensory systems. *Taste* is actually synonymous with *gustation* and refers to the sensations arising from stimulation of the gustatory receptors located within the oropharyngeal mucosa. Throughout this chapter, the terms *taste* and *gustation* are used interchangeably to refer to the gustatory system and its sensations. Taste information arises from stimulation by chemicals dissolved in saliva, which initiate the activation of receptor mechanisms located on specially modified epithelial cells distributed throughout the oral mucosa. Taste transduction initiates depolarization of these receptor cells, which make synaptic contact with first-order fibers of one of several cranial nerves (VII, IX, and X). Fibers from these nerves project to the medulla, into the nucleus of the solitary tract (NST), where second-order projections arise to connect to the pons or thalamus, depending upon the species in question. Pontine neurons project to the thalamus and to various areas of the limbic forebrain involved in the regulation of food and fluid ingestion. Thalamic cells connect to the insular and orbitofrontal cortex. The various populations of taste buds on the anterior and posterior tongue, palate, and laryngeal mucosa have somewhat different sensitivities and project in a topographic, overlapping manner throughout the taste pathway. Cells at all levels of the gustatory system are broadly tuned, typically responding to stimuli that elicit diverse taste qualities. Different neuron types can be identified on the basis of their profiles of sensitivity; they respond best to one of the four basic taste qualities but are typically not specific to any one stimulus. The neural representation of taste quality can be accounted for by the relative patterns of activity evoked across these neuron types. In addition to its obvious role in the perception of salty, sweet, sour, and bitter sensations, taste input is important in regulating a number of visceral reflexes involved in ingestive and digestive functions.

ANATOMY OF THE GUSTATORY SYSTEM

Taste Cells and Taste Buds

The sense of taste serves as a gateway for monitoring and controlling the ingestion of food. It responds to chemical substances in the oral cavity and helps to regulate the interaction between ingestive behavior and the internal milieu. Taste is mediated through chemical stimulation of gustatory receptor cells, which are located within taste buds distributed throughout the oral, pharyngeal, and laryngeal mucosa. Taste buds on the tongue are contained within distinct papillae; those in other areas are distributed across the surface of the epithelium. At the ultrastructural level, at least two kinds of cells can be discerned within the taste bud. They are termed *dark cells* and *light cells* on the basis of their ultrastructural appearance and

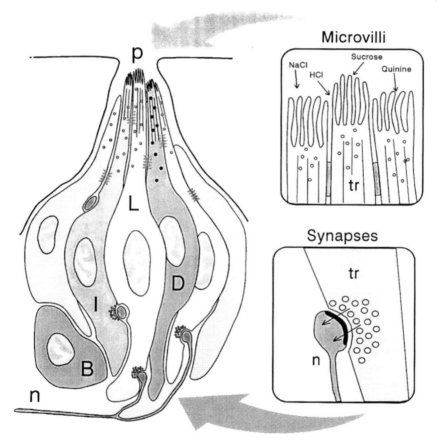

Figure 10–1. Schematic drawing of a mammalian taste bud. This barrel-shaped structure contains different cell types, including basal cells (B), dark cells (D), light cells (L), and intermediate cells (I). These epithelial receptor cells make synaptic contact with distal processes of cranial nerves (n) VII, IX, or X, whose cell bodies lie within the cranial nerve ganglia. The microvilli of the taste receptor cells (tr) project into an opening in the epithelium, the taste pore (p), where they make contact with gustatory stimuli. From Smith and Shipley.[96]

the presence or lack of dense granules in their apical portion[1,2]; *intermediate cells* with characteristics between these extremes are also described.[3] Cells within a taste bud are arranged in a concentric columnar fashion with their apical microvilli projecting toward a pore that opens through the epithelium into the oral cavity (Fig. 10–1); gustatory stimuli interact with membrane receptors and ion channels on these apical microvilli. The base of the taste bud is penetrated by terminal branches of the afferent nerve, which make synaptic contact with the receptor cells.[2,3] All three cell types have been shown to exhibit synaptic specializations.[4] A single nerve fiber may innervate cells in more than one taste bud; each taste bud may be innervated by several different afferent fibers.[5]

Although taste cells are modified epithelial cells, they possess many characteristics of neurons. For example, the neural cell adhesion molecule (NCAM) is expressed on a subset of vallate taste bud cells in the rat[6,7] and mouse[8] and also on the innervating fibers of the glossopharyngeal nerve. Transection of the nerve results in a loss of NCAM expression as the taste buds degenerate,[7,9] indicating that this molecule is expressed by differentiated taste cells and not by the nongustatory epithelial cells. Reinnervation of the vallate papilla following bilateral nerve crush is accompanied by NCAM expression in the nerve, followed by differentiation of the epithelium and the subsequent expression of NCAM in the differentiated taste cells.[9] A similar temporal sequence is seen during taste bud development in the mouse.[8] These studies suggest that NCAM could play some role in the triggering of taste cell differentiation and/or in subsequent axon–taste cell recognition, which occurs during cell turnover. In addition to NCAM, several other molecules, including GAP-43,[10] Thy-1,[11] neuron-specific enolase,[12] keratin-19,[13] and human blood group antigens such as the Lewis[b] blood group epitope[7,14] and blood group determinants A, B, and H,[15] are expressed by taste bud cells. A number of other carbohydrate structures have also been shown on taste buds of several species in lectin-binding studies.[16] The expression of these molecules shows that taste cells display a phenotype characteristic of both neurons and epithelial cells; any of them could play a role in either the structural integrity of the taste bud or in the mediation of axon–taste cell recognition.

Turnover and Replacement of Taste Cells

Taste receptor cells arise continually from an underlying layer of basal epithelial cells. It is not clear, primarily because the cells are in a constant state of turnover, whether the cell types identifiable on ultrastructural grounds are different cell types or a single type at different stages of differentiation.[1,3] In the rat, the life span of a taste cell in a fungiform papilla is approximately 10 days.[17] Dark cells in the rat vallate papilla have a life span of about 9 days; light cells appear to turn over more slowly, suggesting that they are a cell type distinct from the dark cells.[1] The afferent nerve maintains a trophic influence over the taste buds, which will degenerate if their nerve supply is removed.[18,19] This degeneration is accompanied by a loss of expression of the molecules expressed by differentiated taste cells.[7,9,15,] Although the innervation by gustatory nerve fibers is necessary to maintain the structural and biochemical integrity of the taste bud, the gustatory sensitivities of the receptor cells appear to be determined by the epithelium itself and not by the innervating nerve.[20] That is, when the chorda tympani and glossopharyngeal nerves are cross-reinnervated, the gustatory sensitivities of the anterior and posterior tongue are not altered. Interestingly, the several branches of a chorda tympani nerve fiber that innervate different fungiform papillae have been shown to have similar profiles of sensitivity.[21] Combined with the fact that the sensitivity of a given receptor field appears to be determined by the epithelium, this strongly suggests that during cell turnover the nerve fibers are guided to make contact with particular types of receptor cells.

Distribution and Innervation of Taste Buds

Taste buds are found on the anterior portion of the tongue in fungiform papillae and in circumvallate and foliate papillae on the posterior tongue.[22-25] There are also taste buds on the soft palate, pharynx, epiglottis, and upper third of the esophagus.[25-27] The distribution of taste buds on the human tongue and within the oral cavity is shown schematically in Figure 10–2. In humans, the 200 to 300 fungiform papillae on the anterior portion of the tongue contain approximately 1600 taste buds, although there is considerable variation among individuals.[28,29] The 8 to 12 circumvallate papillae contain about 250 taste buds each, for a total of nearly 3000 taste buds, and the foliate papillae have about 1300 taste buds.[23,30] Although taste buds have been described on the soft palate of human adults only in biopsy material and only in very small numbers,[27] human infants are reported to have about 2600 taste buds in the pharynx and larynx and on the soft palate.[31] These data on the numbers of human taste buds, however, are derived from very few studies. On the tongue of adult rhesus monkeys (fungiform, circumvallate, and foliate papillae), there are approximately 8000 to 10000 taste buds, which are maintained well into old age.[26] Most electrophysiological studies of taste have employed stimulation of the fungiform papillae, even though the majority of taste buds in all mammalian species studied are located in other areas.

Taste receptors are innervated by branches of the seventh (facial), ninth (glossopharyngeal), and tenth (vagus) cranial nerves. Taste buds in the fungiform papillae on the anterior portion of the tongue are innervated by the chorda tympani branch of the facial nerve, and those on the soft palate are innervated by its greater superficial petrosal branch, via the lesser palatine nerve. The cell bodies of these afferent fibers are located in the geniculate ganglion and project centrally into the rostral pole of the NST.[32,33] Circumvallate and most foliate taste buds are supplied by the lingual-tonsillar branch of the glossopharyngeal nerve, although the most rostral foliate taste buds are innervated by the chorda tympani nerve. Afferent fibers of the glossopharyngeal nerve project through the inferior glossopharyngeal (petrosal) ganglion to the NST just caudal to, but overlapping with, the facial nerve termination.[32,33] The pharyngeal branch of the glossopharyngeal nerve innervates taste buds in the nasopharynx.[34] Taste buds on the epiglottis, aryepiglottal folds, and esophagus are innervated by the internal portion of the superior laryngeal branch of the vagus nerve. Afferent fibers from the superior laryngeal nerve project via their cell bodies in the inferior vagal (nodose)

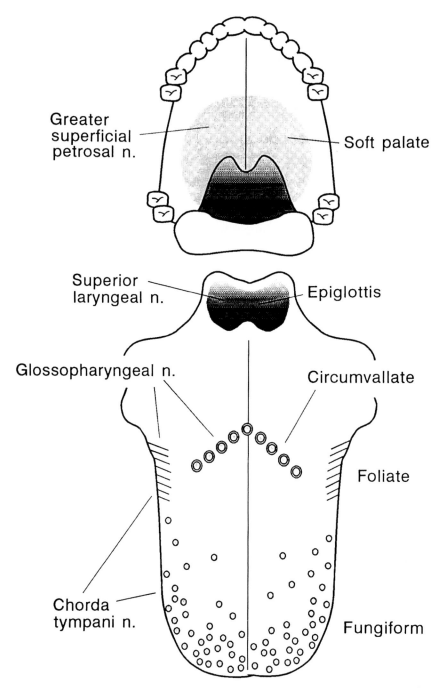

Figure 10–2. Diagram of the human oral cavity, showing the distribution of various taste bud populations, which are found on the anterior tongue (fungiform papillae), posterior tongue (circumvallate and foliate papillae), the soft palate, and the laryngeal surface of the epiglottis. These taste buds are innervated by several different cranial nerves, which are branches of the seventh, ninth, and tenth nerves (see text). From Smith and Shipley.[96]

ganglion to the NST caudal to, but overlapping with, the glossopharyngeal nerve termination.[32,33,35]

Central Gustatory Pathways

From the NST, secondary gustatory fibers arise to project rostrally more or less parallel to the projection of general visceral sensation, which arises from the more caudal aspects of the solitary nucleus.[36] In most mammalian species, there is a third-order projection into the parabrachial nuclei (PbN) in the pons, from which fibers arise to project ipsilaterally via a classic sensory path to the parvicellular division of the ventroposteromedial nucleus of the thalamus and then to the agranular insular cortex (Fig. 10–3). In addition to this thalamocortical projection, fibers also travel from the pons into areas of the ventral forebrain involved in feeding and autonomic regulation, including the lateral hypothalamus, central nucleus of the amygdala, and the bed nucleus of the stria terminalis.[37] In primates, however, taste fibers bypass the pontine relay and project ipsilaterally through the central tegmental tract directly to the VPMpc in the thalamus.[38] The VPMpc projects in the monkey to the insular and opercular cortex and to area 3b on the lateral convexity of the precentral gyrus.[39] What may be a secondary cortical area, receiving input from the anterior insula,

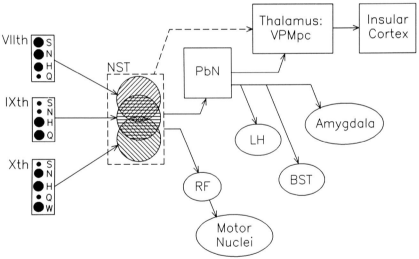

Figure 10–3. Schematic diagram of the mammalian gustatory afferent pathway. Peripheral fibers with different degrees of responsiveness to sucrose (S), NaCl (N), HCl (H), quinine hydrochloride (Q), or water (W) project into the rostral pole of the nucleus of the solitary tract (NST). Fibers of the seventh, ninth, and tenth nerves project in an organized overlapping termination within the rostral portion of the NST. Second-order cells project into the parabrachial nuclei (PbN) of the pons, where a classic lemniscal pathway proceeds to the parvicellular division of the ventroposteromedial nucleus (VPMpc) of the thalamus and then to the insular cortex. Another projection arises within the PbN to connect to areas of the ventral forebrain (lateral hypothalamus, LH; bed nucleus of the stria terminalis, BST; amygdala) involved in the control of feeding and in autonomic regulation. Cells of the NST also make reflex connections via the reticular formation (RF) with cranial motor nuclei that control muscles involved in facial expression, licking, chewing, and swallowing. In primates, cells of the NST bypass the PbN and project directly to the thalamus (dashed line). From Smith and Shipley.[97]

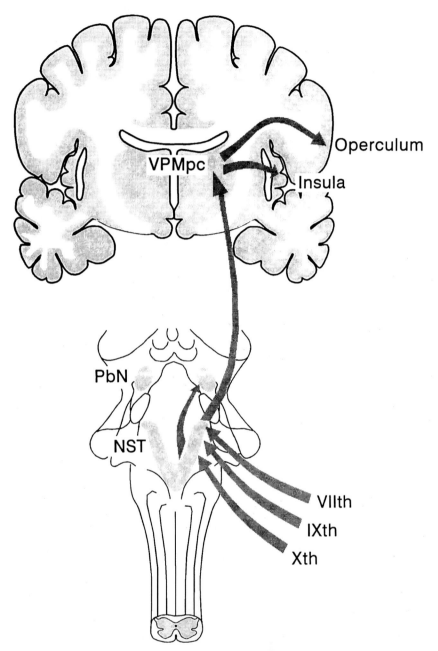

Figure 10–4.   Drawing of the probable gustatory afferent pathway in humans. Fibers of the seventh, ninth, and tenth nerve project into the nucleus of the solitary tract (NST), from which there is a direct projection to the ipsilateral thalamus, to the parvicellular division of the ventropostermedial nucleus (VPMpc). From the thalamus, fibers project into the insular cortex and frontal operculum (see text). Visceral afferent fibers from the caudal NST project to the parabrachial nuclei (PbN), which also receives gustatory projections in nonprimates. In addition to this classical sensory projection, there are known to be numerous subcortical projections of the gustatory nuclei in a variety of species (see Fig. 10–3). From Smith and Shipley.[96]

is found within the posterior orbitofrontal cortex.[40] None of these cortical areas, however, are purely gustatory, since they all contain neurons responsive to other sensory modalities such as touch and temperature.[41] There are, unfortunately, few clear data on the anatomy of these projections in humans,[42] although a recent careful review of the clinical literature suggests that the human gustatory system is quite similar to that of the Old World monkey.[41] A schematic of the probable gustatory projections in humans, based on data from other primates, is shown in Figure 10–4.

TASTE PHYSIOLOGY

Receptor Physiology and Transduction

The mechanisms of taste transduction are currently receiving a great deal of attention.[43–45] Although the nature of gustatory transduction is not completely understood, a coherent story is beginning to emerge from studies of isolated amphibian and mammalian receptor cells (Fig. 10–5). Sour taste is produced by acids (H^+), and studies have now shown that voltage-dependent K^+ channels restricted to the apical membrane of the taste receptor cell are involved in the transduction of acids.[46] Acidic sour stimuli block the K^+ channels on the apical membrane, which results in a direct depolarization of the cell, leading to excitation of the basolateral membrane.

There is a wealth of evidence that voltage-independent Na^+ channels on the apical membranes of taste cells in mammals and frogs partially mediate the transduction of Na^+ salt taste.[43–45] The presence of these passive channels in the apical membrane allows the flow of inward current in the presence of a Na^+ stimulus, which depolarizes the cell. Treatment of the tongue surface with amiloride, an epithelial Na^+-channel blocker, reduces the Na^+ salt–evoked nerve activity in chorda tympani fibers and the short-circuit current across lingual epithelium[44,47] and blocks whole-cell passive Na^+ currents in isolated frog taste cells.[43] An additional non-amiloride-sensitive pathway has also been implicated in the transduction of NaCl, which involves current flow through a paracellular pathway to allow stimulation by Na^+ of basolateral ion channels.[48]

These transduction events (for acids and Na^+ salts) do not require the existence of specific membrane receptors, but depend upon the direct action of the stimulating ions on apical membrane channels. Transduction of sweet- and bitter-tasting stimuli is not as well understood, but appears to involve specific membrane receptors linked to second messenger systems.[44,45,49] There is some evidence that there may be more than one type of receptor for sweet stimuli,[50,51] and multiple receptor mechanisms are also suggested for bitter-tasting compounds.[14,52–54] Although there have not been extensive systematic studies of the profiles of sensitivity of gustatory receptor cells, available evidence (mostly from microelectrode recordings in situ) suggests that taste cells in amphibians and mammals may be broadly sensitive to stimuli representing different taste qualities, responding often to two or more of the four basic taste stimuli: NaCl, sucrose, quinine, and acid.[51,55–57] However, there are no data

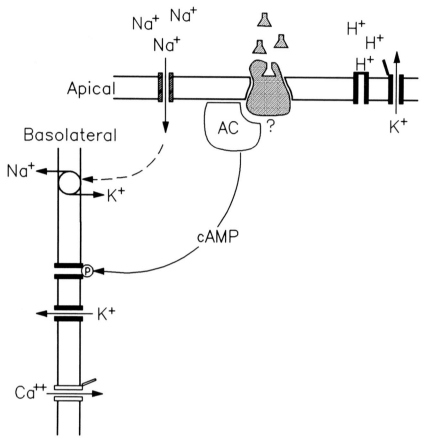

Figure 10–5. Schematic representation of taste transduction mechanisms. Transduction of the taste of acids involves H$^+$ blockage of voltage-dependent K$^+$ channels on the apical membrane of the taste receptor cell. Sodium salt (Na$^+$) transduction involves the passage of Na$^+$ into the receptor cell through passive, amiloride-blockable ion channels on the apical membrane, and also by basolateral ion channels (not shown). Sodium balance is then restored through a (Na$^+$, K$^+$)-ATPase on the basolateral membrane. The transduction of sweet- and bitter-tasting stimuli probably involves a number of different receptor proteins on the apical membrane, which bind specifically to these substances. It has been shown that transduction of sweet substances involves receptor-mediated stimulation of adenylate cyclase, which leads to a closing of voltage-independent K$^+$ channels on the basolateral membrane by a cyclic adenosine monophosphate–dependent phosphorylation. However, the nature of the mediation between receptor binding and adenylate cyclase activity has not been revealed. All of these mechanisms lead to depolarization and influx of Ca^{2+} through voltage-dependent Ca^{2+} channels. From Smith and Shipley.[97]

obtained from newer whole-cell recording methods that adequately address the issue of the multiple sensitivity of taste receptor cells.

Neural Representation of Taste Quality

Many textbooks of physiology show a diagram of the human tongue that suggests that saltiness and sweetness are detected at the tip, sour on the sides, and bitter on the back of the tongue. Although there are slight differences in the absolute threshold for different taste qualities in different regions of the human tongue and palate,[58] all taste qualities (salty, sour, sweet, and bitter) can be perceived by stimulation of each of the taste bud populations. Physiological studies in rodents, on the other hand, suggest some fairly striking differences between the various taste bud populations in their response to different tastants.[59] Fibers of the seventh nerve are much more responsive to sweet and salty stimuli, whereas those of the ninth nerve are relatively more responsive to sour and bitter substances. These differences may relate to different functional roles for the separate taste bud populations (see below).

The perception of saltiness, sweetness, sourness, or bitterness emerges from neural activity within the central nervous system. These psychological concepts are used by humans to describe the sensations arising from stimulation of gustatory receptors by a variety of chemical stimuli. Information necessary for these perceptions is carried to the brain by the activity in peripheral taste nerves. The response profiles of peripheral gustatory nerve fibers reflect the way in which the sensitivities of taste receptor cells are distributed among these first-order neurons. Transduction of specific chemical stimuli (eg, sodium ions, protons, sugars, alkaloids) by taste receptors gives rise to activity in several types of afferent nerve fibers. Understanding the neural coding of taste information must begin with knowledge about how chemical sensitivities, represented by specific transduction mechanisms, are distributed and organized among peripheral and central gustatory neurons. Taste receptor mechanisms are distributed across the several different subpopulations of taste buds, which are located on different regions of the tongue and oral epithelium and are innervated by one of several cranial nerves. The taste-responsive fibers in these nerves can be classified on the basis of their response spectra, and fibers in the seventh, ninth, and tenth cranial nerves have very different profiles of sensitivity, which reflect differential input from several taste transduction mechanisms in the periphery.[59]

Most of what is known about the neurophysiology of the mammalian gustatory pathway has been derived from studies on the input from the anterior portion of the tongue. Like individual receptor cells, single fibers in the chorda tympani nerve typically respond to more than one taste quality.[60,61] Individual second- and third-order gustatory neurons in the nucleus of the solitary tract and parabrachial nuclei that respond to anterior tongue stimulation are similarly broadly tuned across taste quality.[62–66] The responses of fibers in the glossopharyngeal nerve also demonstrate a lack of stimulus specificity.[67,68]

However, even in the face of this broad tuning, individual gustatory cells can be categorized on the basis of similarities in their response profiles and their predominant sensitivities, ie, they appear to fall into functional groups, which can be identified by their "best" stimulus.[60,61,63,68] Attempts to understand the neural processing of taste quality information have relied heavily on this "best-stimulus" classification (ie, sucrose-best, NaCl-best, etc), which implies the existence of four basic taste qualities: salty, sour, sweet, and bitter.[59]

Inputs from the separate peripheral taste fields project into the NST, where it has been shown that the cells are somewhat more broadly tuned than peripheral fibers[64] and often receive converging input from separate peripheral fields.[69,70] The response of a broadly tuned cell in the hamster NST is shown in Figure 10–6. Of the

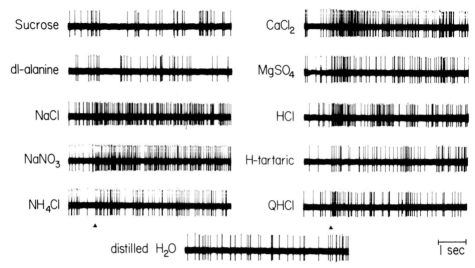

Figure 10–6.　Responses of a neuron in the nucleus of the solitary tract (NST) of the hamster to several taste stimuli applied to the fungiform papillae. The arrow indicates the onset of the response, of which about 5 seconds is shown, preceded by about 1 second of response to distilled water. The concentrations of the stimuli are those that produce a half-maximal response to these chemicals in the hamster's chorda tympani nerve. This cell shows an excitatory response to all of these stimuli except sucrose and dl-alanine, which produce an inhibition of ongoing activity. From Smith et al.[98]

four prototypical stimuli (sucrose, NaCl, HCl, and quinine-HCl [QHCl]), this cell responds best to HCl, but it also responds to NaCl and QHCl; it is inhibited by sucrose. Further increases in breadth of responsiveness have been shown in the third-order cells of the PbN.[66]

　　Taste fibers in peripheral gustatory nerves and cells in central taste nuclei are fairly broadly tuned across stimulus quality, and it is generally agreed that the representation of taste quality involves a comparison of activity across taste fibers.[59,71–73] This across-fiber pattern theory of quality coding was first suggested by Pfaffmann[74,75] when it was apparent that the earliest recordings of taste afferent fiber activity indicated a lack of stimulus specificity. The across-neuron patterns of activity can be quantified by calculating the correlations among the responses evoked by a series of chemical stimuli across a sample of taste neurons.[72] These correlations can then be subjected to a multivariate analysis to create a "taste space" that represents the neurophysiological similarities and dissimilarities among the stimuli. The across-neuron taste space for 18 stimuli generated from the responses of cells in the hamster PbN are shown in Figure 10–7, where it is evident that stimuli with similar taste are grouped together. The similarities and differences in the patterns of activity evoked by these stimuli provide a basis for their discrimination. By relating the responses of the various fiber types (ie, sucrose-best, NaCl-best, etc) to these across-neuron patterns, it has been shown that a comparison of activity across the various gustatory fiber types appears to be necessary for the neural discrimination among stimuli of different taste quality.[59,73,76] Based on the specific effects of the ion channel blocker, amiloride, there have been recent suggestions that saltiness may be coded specifically by activity in a single class of NaCl-best cells.[77,78] However, these results do not exclude the distinct possibility that activity in NaCl-best cells is a necessary but not sufficient code for the salty taste, ie, the discrimination of salts from other stimuli requires a comparison of the activities in several fiber types.[59,73,76] This conclusion is further complicated by recent human psychophysical data showing no effect of amiloride on the perception of the salty quality.[79]

Quite apart from the issue of quality coding mechanisms is the interesting question of how, in light of the broad tuning of taste receptor cells and first-order fibers, any kind of code for quality can be maintained in the face of the constant turnover of receptor cells.[1,19] Electrophysiological work on cross-regenerated taste fibers—in which the chorda tympani and glossopharyngeal nerves were cut, crossed, and allowed to reinnervate the tongue[20]—demonstrated that the tongue epithelium

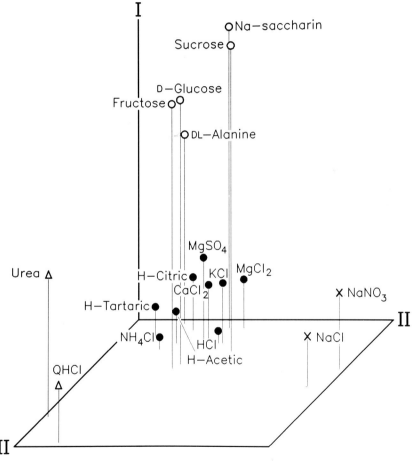

Figure 10–7. Three-dimensional "taste space" showing the similarities and differences in the across-neuron patterns evoked by 18 stimuli delivered to the anterior tongue of the hamster. This space was derived from multidimensional scaling (KYST, Bell Laboratories) of the across-neuron correlations among these stimuli recorded from neurons in the parabrachial nuclei (PbN) of the hamster. Close proximity within this space indicates a high degree of correlation between the across-neuron patterns elicited by these stimuli. Four groups of stimuli are indicated by different symbols: sweeteners, sodium salts, nonsodium salts and acids, and bitter-tasting stimuli. From Smith et al.[76]

determined the relative sensitivities of the two nerves. This was somewhat surprising, since the nerves themselves have a trophic influence over the taste buds.[18,19] Nevertheless, the representation of taste quality in gustatory nerve fibers must somehow remain constant during the continual turnover of taste receptor cells and their reinnervation by fibers of the peripheral nerve.

Gustatory-Mediated Behavior

Information arising from the various gustatory nerves projects into the NST of the medulla. Here second-order neurons give rise to ascending projections to the PbN of the pons (in nonprimates), which in turn sends projections to the thalamus and insular cortex and to widespread areas of the limbic forebrain. There are also numerous connections of NST neurons to the oral motor nuclei via interneurons in the reticular formation.[80,81] These anatomical relationships; the differential sensitivities of the seventh, ninth, and tenth nerves; and the contribution of gustatory afferent input to taste-mediated behaviors are summarized in the schematic diagram of Figure 10–8. Whereas taste physiologists have focused largely on the role of gustatory afferent fibers and central neurons in taste quality perception, there are a number of taste-mediated behaviors ranging from tongue movements to salivation to preabsorptive insulin release that have their neuronal substrate within the brainstem.[82–84] Input from the various gustatory nerves contributes differentially to these diverse taste-mediated behaviors.

One way of viewing taste is as the rostral component of a visceral afferent system, which includes gustatory, respiratory, cardiovascular, and gastrointestinal functions.[36] Taste buds innervated by the seventh, ninth, and tenth nerves contribute differentially to this visceral continuum, with seventh nerve fibers responsive primarily to preferred stimuli like sucrose and NaCl,[60,61,85] ninth nerve fibers most sensitive to aversive stimuli like HCl and QHCl,[67,68] and tenth nerve fibers responsive to stimuli that deviate from the normal pH and ionic milieu of the larynx.[86] Sucrose, for example, predominantly stimulates fibers of the seventh nerve. These fibers project into the NST, where cells that are sucrose-best are more broadly tuned than chorda tympani fibers in the hamster; many of these cells also respond to NaCl and to HCl, but are often inhibited by QHCl.[64] Ultimately, the output of these second-order neurons ascends to the forebrain to give rise to the perception of sweetness (Fig. 10–8). At the same time, these cells provide input to motor systems that drive the ingestive components of feeding behavior, including rhythmic mouth movements, tongue protrusions, lateral tongue protrusions, salivary secretion, insulin release, and swallowing.[82,84,87] Conversely, QHCl predominantly stimulates fibers of the ninth nerve. These fibers project into the NST, where they drive cells that are also responsive to HCl and NaCl but not to sucrose.[69] Quinine-sensitive cells of the NST send ascending projections to the forebrain to give rise to sensations of bitterness (Fig. 10–8), but they also provide input to motor systems that drive protective behaviors in rodents like gaping, chin rubbing, forelimb flailing, locomotion, and fluid rejection.[82,87]

In the rat, the number of gapes elicited by quinine stimulation is reduced by almost one half after bilateral transection of the ninth nerve.[88] Sucrose and

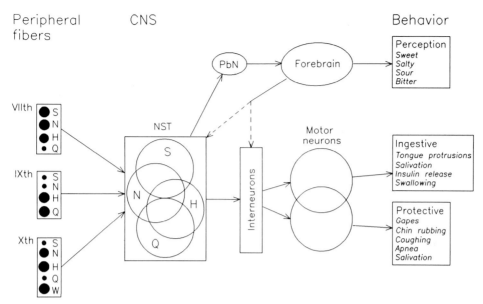

Figure 10–8. Schematic diagram of the chemosensory inputs of three cranial nerves to the taste-responsive portion of the nucleus of the solitary tract (NST) and their putative role in taste-mediated behaviors. The size of the filled circles for each of the peripheral nerves (VII, IX, and X) depicts the relative responsiveness of these nerves to sucrose (S), NaCl (N), HCl (H), QHCl (Q), and water (W). Sensitivities of NST cells are largely overlapping, with each cell type somewhat responsive to two or three of the basic stimuli. Sucrose and QHCl stimulate few of the same NST cells, however. Output from the NST ascends in the classic taste pathway to give rise to perceptions of sweetness, saltiness, sourness, and bitterness and to hedonic tone (not depicted). Local reflex circuits within the brainstem control ingestive and protective responses evoked by taste stimulation. Behavioral data suggest that both ingestive and protective responses can be triggered in parallel, depending upon the quality of the stimulus. From Smith and Frank.[59]

quinine produce very different patterns of ingestive and protective taste reactivity,[82] and a combination of these behaviors can be triggered by mixtures of sucrose and quinine.[89]

Similarly, in the hamster the patterns of taste reactivity to sucrose and quinine are quite different, whereas sodium salts and acids produce patterns consisting of combinations of both ingestive and protective behaviors.[87] The taste quality of the stimulus is very directly related to the specific pattern of taste reactivity that is elicited in the hamster.

The superior laryngeal branch of the tenth nerve is involved in swallowing, airway protection, and a number of other visceral reflexes. Respiratory apnea is produced by laryngeal stimulation with water[90,91] and chemosensory fibers of the rat superior laryngeal nerve have been shown to mediate diuresis in response to stimulation of the laryngeal mucosa with water.[92] Thus, chemoreceptive fibers of the tenth nerve are more involved in visceral functions than in taste quality perception.[86]

Besides its obvious role in controlling ingestive behavior, taste also triggers a number of metabolic responses, including salivary, gastric, and pancreatic secretions,[84] although the specific contributions of particular cranial nerves to these responses are not well understood. Thus, in addition to their mediation of gustatory sensation, taste buds may have a number of roles related to gustatory-visceral regulation, depending upon their peripheral distribution and innervation.

The gustatory system readily adapts to constant stimulation. For example, flowing sodium chloride over the tongue for 60 seconds results in a complete disappearance of the salty sensation.[93] Taste receptors are particularly sensitive to changing stimuli, ie, to the rate of concentration change.[94] Indeed, the natural course of events in tasting involves intermittent and rapid contact of the various taste buds as stimuli are moved through the oral cavity during biting, chewing, and swallowing.[95] Because of the rapid adaptation to constant stimulation, care must be taken in gustatory testing procedures to guard against changes in sensitivity caused by prolonged and repeated stimulation.

Acknowledgments—Many of the data reported in this chapter were acquired under the support of NIDCD grants DC00353, DC00066, and DC00347; the author was partially supported by these awards during the preparation of the manuscript.

REFERENCES

1. Farbman AI: Renewal of taste bud cells in rat circumvallate papillae. Cell Tissue Kinetics 13:349–357, 1980.
2. Murray RG: Ultrastructure of taste receptors. In: Beidler LM (ed): Handbook of Sensory Physiology. Vol. IV: Chemical Senses. Part 2: Taste. Berlin: Springer-Verlag, 1971.
3. Kinnamon JC: Organization and innervation of taste buds. In: Finger T, Silver WL (eds): Neurobiology of Taste and Smell. New York: John Wiley & Sons, 1987.
4. Kinnamon JC, Sherman TA, Roper SD: Ultrastructure of mouse vallate taste buds: III. Patterns of synaptic connectivity. J Comp Neurol 270:1–10, 1988.
5. Beidler LM: Innervation of rat fungiform papilla. In: Pfaffmann C (ed): Olfaction and Taste III. New York: Rockefeller University Press, 1970.
6. Nelson GM, Finger TE: Immunolocalization of different forms of neural cell adhesion molecule (NCAM) in rat taste buds. J Comp Neurol 336:507–516, 1993.
7. Smith DV, Akeson RA, Shipley MT: NCAM expression by subsets of taste cells is dependent upon innervation. J Comp Neurol 336:493–506, 1993.
8. Nolte C, Martini R: Immunocytochemical localization of the L1 and N-CAM cell adhesion molecules and their shared carbohydrate epitope L2/HNK-1 in the developing and differentiated gustatory papillae of the mouse tongue. J Neurocytol 21:19–33, 1992.
9. Smith DV, Klevitsky R, Akeson RA, Shipley MT: Expression of the neural cell adhesion molecule (NCAM) and polysialic acid during taste bud degeneration and regeneration. J Comp Neurol 347:187–196, 1994.
10. Finger T, Benowitz L, Pfenninger K: Immunocytochemical localization of neuronal growth-related membrane proteins in rodent taste buds. Chem Senses 13:688, 1988.
11. Lasher RS, Erickson PF: Localization of Thy-l-like immunoreactivity in the rat and mouse taste bud. Neurosci Abstr 15:891, 1989.
12. Yoshie S, Wakasugi C, Teraki Y, Iwanaga T, Fujita T: Immunocytochemical localizations of neuron-specific proteins in the taste bud of the guinea pig. Arch Histol Cytol 51:379–384, 1988.
13. Oakley B: Neuronal-epithelial interactions in mammalian gustatory epithelium. In: Rubel E (ed): Regeneration of Vertebrate Sensory Receptor Cells. Chichester, UK: John Wiley, 1991.
14. Akabas MH, Dodd J, Al-Awqati Q: A bitter substance induces a rise in intracellular calcium in a subpopulation of rat taste cells. Science 242:1047–1050, 1988.
15. Smith DV, Klevitsky R, Akeson RA, Shipley MT: Taste bud expression of human blood group antigens. J Comp Neurol 343:130–142, 1994.
16. Witt M, Reutter K: Lectin histochemistry on mucous substances of the taste buds and adjacent epithelia of different vertebrates. Histochemistry 88:453–461, 1988.
17. Beidler LM, Smallman RL: Renewal of cells within taste buds. J Cell Biol 27:263–272, 1965.
18. Guth L: Degeneration and regeneration of taste buds. In: Beidler LM (ed): Handbook of Sensory Physiology. Vol. IV: Chemical Senses. Part 2: Taste. Berlin: Springer-Verlag, 1971.

19. Oakley B: Trophic competence in mammalian gustation. In: Pfaff D (ed): Taste, Olfaction, and the Central Nervous System. New York: Rockefeller University Press, 1985.
20. Oakley B: Altered temperature and taste responses from cross-regenerated sensory nerves in the rat's tongue. J Physiol 188:353–371, 1967.
21. Oakley B: Receptive fields of cat taste fibers. Chem Senses Flavor 1:431–442, 1975.
22. Arvidson K: Location and variation in number of taste buds in human fungiform papillae. Scand J Dent Res 87:435–442, 1979.
23. Bradley RM: Tongue topography. In: Beidler LM (ed): Handbook of Sensory Physiology. Vol. IV: Chemical Senses. Part 2: Taste. Berlin: Springer-Verlag, 1971.
24. Miller IJ Jr: Gustatory receptors of the palate. In: Katsuki Y, Sato M, Takagi S, Oomura Y (eds): Food Intake and Chemical Senses. Tokyo: University of Tokyo Press, 1977.
25. Miller IJ Jr, Smith DV: Quantitative taste bud distribution in the hamster. Physiol Behav 32:275–285, 1984.
26. Bradley RM, Stedman HM, Mistretta CM: Age does not affect numbers of taste buds and papillae in adult rhesus monkeys. Anat Record 212:246–249, 1985.
27. Nilsson B: The occurrence of taste buds in the palate of human adults as evidenced by light microscopy. Acta Odontol Scand 37: 253–258, 1979.
28. Miller IJ Jr: Variation in human fungiform taste bud densities among regions and subjects. Anat Record 216:474–482, 1986.
29. Miller IJ Jr, Bartoshuk LM: Taste perception, taste bud distribution, and spatial relationships. In: Getchell TV, Bartoshuk LM, Doty RL, Snow JB Jr (eds): Smell and Taste in Health and Disease. New York: Raven Press, 1991.
30. Mochizuki Y: Studies on the papillae foliatae of Japanese. 2. The number of taste buds. Okajimas Folia Anat Japan 18:355–369, 1939.
31. Lalonde ER, Eglitis JA: Number and distribution of taste buds on the epiglottis, pharynx, larynx, soft palate and uvula in a newborn human. Anat Record 140:91–93, 1961.
32. Contreras RJ, Beckstead RM, Norgren R: The central projections of the trigeminal, facial, glossopharyngeal and vagus nerves: An autoradiographic study in the rat J Autonomic Nerv System 6:303–322, 1982.
33. Hamilton RB, Norgren R: Central projections of gustatory nerves in the rat J Comp Neurol 222:560–577, 1984.
34. Travers SP, Nicklas K: Taste bud distribution in the rat pharynx and larynx. Anat Record 227:373–379, 1990.
35. Hanamori T, Smith DV: Central projections of the hamster superior laryngeal nerve. Brain Res Bull 16:271–279, 1986.
36. Norgren R: Taste and the autonomic nervous system. Chem Senses 10:143–161, 1985.
37. Norgren R, Leonard CM: Ascending central gustatory pathways. J Comp Neurol 150:217–238, 1973.
38. Beckstead R, Morse J, Norgren R: The nucleus of the solitary tract in the monkey: Projections to the thalamus and brain stem nuclei. J Comp Neurol 190:259–282, 1980.
39. Pritchard TC, Hamilton RB, Morse J, Norgren R: Projections from thalamic gustatory and lingual areas in the monkey (*Macaca fascicularis*). J Comp Neurol 244:213–228, 1986.
40. Thorpe SJ, Rolls ET, Maddison S: The orbitofrontal cortex: Neuronal activity in the behaving monkey. Exp Brain Res 49:93–115, 1983.
41. Pritchard TC: The primate gustatory system. In: Getchell TV, Bartoshuk LM, Doty RL, Snow JB Jr (eds): Smell and Taste in Health and Disease. New York: Raven Press, 1991.
42. Norgren R: Gustatory system. In: Paxinos G (ed): The Human Nervous System. San Diego: Academic Press, 1990.
43. Avenet P, Lindemann B: Amiloride-blockable sodium currents in isolated taste receptor cells. J Membrane Biol 105:245–255, 1988.
44. Heck GL, Mierson S, DeSimone JA: Salt taste transduction occurs through an amiloride-sensitive sodium transport pathway. Science 223:403–405, 1984.
45. Kinnamon SC: Taste transduction: A diversity of mechanisms. Trends Neurosci 11:491–496, 1988.
46. Kinnamon SC, Dionne VE, Beam KG: Apical localization of K^+ channels in taste cells provides the basis for sour taste transduction. Proc Natl Acad Sci USA 85:7023–7027, 1988.
47. Heck GL, Persaud KC, DeSimone JA: Direct measurement of translingual epithelial NaCl and KCl currents during the chorda tympani taste response. Biophys J 55:843–857, 1989.
48. Ye Q, Heck GL, DeSimone JA: Voltage dependence of the rat chorda tympani response to Na^+ salts: Implications for the functional organization of taste receptor cells. J Neurophysiol 70:167–178, 1993.
49. Tonosaki K, Funakoshi M: Cyclic nucleotides may mediate taste transduction. Nature 331:354–356, 1988.

50. Jakinovich W Jr, Sugarman D: Peripheral mechanisms of mammalian sweet taste. In: Cagan RH (ed): Neural Mechanisms of Taste. Boca Raton, FL: CRC Press, 1989.

51. Tonosaki K, Funakoshi M: Intracellular taste cell responses of mouse. Comp Biochem Physiol 78A:651–656, 1984.

52. Herness MS, Pfaffmann C: Generalization of conditioned taste aversions in hamsters: Evidence for multiple bitter receptor sites. Chem Senses 11:347–360, 1986.

53. McBurney DH, Smith DV, Shick TR: Gustatory cross adaptation: Sourness and bitterness. Percep Psychophys 11:228–232, 1972.

54. Okada Y, Miyamoto T, Sato T: Ionic mechanism of generation of receptor potential in response to quinine in frog taste cell. Brain Res 450:295–302, 1988.

55. Ozeki M, Sato M: Responses of gustatory cells in the tongue of rat to stimuli representing four taste qualities. Comp Biochem Physiol 41A:391–407, 1972.

56. Sato T: Multiple sensitivity of single taste cells of the frog tongue to four basic taste stimuli. J Cell Physiol 80:207–218, 1972.

57. Sato T: Recent advances in the physiology of taste cells. Prog Neurobiol 14:25–67, 1980.

58. Collings VB: Human taste response as a function of locus of stimulation on the tongue and soft palate. Percept Psychophys 16:169–174, 1974.

59. Smith DV, Frank ME: Sensory coding by peripheral taste fibers. In: Simon SA, Roper SD (eds): Mechanisms of Taste Transduction. Boca Raton, FL: CRC Press, 1993.

60. Frank M: An analysis of hamster afferent taste nerve response functions. J Gen Physiol 61:588–618, 1973.

61. Frank ME, Bieber SL, Smith DV: The organization of taste sensibilities in hamster chorda tympani nerve fibers. J Gen Physiol 91:861–896, 1988.

62. Smith DV: Brainstem processing of gustatory information. In: Pfaff D (ed): Taste, Olfaction, and the Central Nervous System. New York: Rockefeller University Press, 1985.

63. Smith DV, Van Buskirk RL, Travers JB, Bieber SL: Gustatory neuron types in hamster brain stem. J Neurophysiol 50:522–540, 1983.

64. Travers JB, Smith DV: Gustatory sensitivities in neurons of the hamster nucleus tractus solitarius. Sensory Processes 3:1–26, 1979.

65. Travers SP, Smith DV: Responsiveness of neurons in the hamster parabrachial nuclei to taste mixtures. J Gen Physiol 84:221–250, 1984.

66. Van Buskirk RL, Smith DV: Taste sensitivity of hamster parabrachial pontine neurons. J Neurophysiol 45:144–171, 1981.

67. Frank ME: Taste-responsive neurons of the glossopharyngeal nerve of the rat. J Neurophysiol 65:1452–1463, 1991.

68. Hanamori T, Miller IJ Jr, Smith DV: Gustatory responsiveness of fibers in the hamster glossopharyngeal nerve. J Neurophysiol 60:478–498, 1988.

69. Sweazey RD, Smith DV: Convergence onto hamster medullary taste neurons. Brain Res 408:173–184, 1987.

70. Travers SP, Norgren R: Coding the sweet taste in the nucleus of the solitary tract: Differential roles for anterior tongue and nasoincisor duct gustatory receptors in the rat. J Neurophysiol 65:1372–1380, 1991.

71. Erickson RP: On the neural basis of behavior. Am Scientist 72:233–241, 1984.

72. Erickson RP, Doetsch GS, Marshall DA: The gustatory neural response function. J Gen Physiol 49:247–263, 1965.

73. Smith DV, Vogt MB: The neural code and integrative processes of taste. In: Bartoshuk LM, Beauchamp GK (eds): Handbook of Perception and Cognition: Tasting and Smelling. New York: Academic Press, in press.

74. Pfaffmann C: Gustatory nerve impulses in rat, cat and rabbit J Neurophysiol 18:429–440, 1955.

75. Pfaffmann C: The afferent code for sensory quality. Am Psychol 14:226–232, 1959.

76. Smith DV, Van Buskirk RL, Travers JB, Bieber SL: Coding of taste stimuli by hamster brain stem neurons. J Neurophysiol 50:541–558, 1983.

77. Scott TR, Giza BK: Coding channels in the taste system of the rat. Science 249:1585–1587, 1990.

78. Scott TR, Plata-Salaman C: Coding of taste quality. In: Getchell TV, Bartoshuk LM, Doty RL, Snow JB Jr (eds): Smell and Taste in Health and Disease. New York: Raven Press, 1991.

79. Ossebaard CA, Smith DV: Effect of amiloride on the taste of NaCl, Na-gluconate and KCl in humans: Implications for Na$^+$ receptor mechanisms. Chem Senses 20:37–46, 1995.

80. Travers JB: Organization and projections of the orofacial motor nuclei. In: Paxinos G (ed): The Rat Nervous System. Vol. 2. Hindbrain and Spinal Cord. Sydney: Academic Press, 1985.

81. Travers JB, Norgren R: Afferent projections to the oral motor nuclei in the rat. J Comp Neurol 220:280–298, 1983.

82. Grill HJ, Norgren R: The taste reactivity test. I. Mimetic responses to gustatory stimuli in neurologically normal rats. Brain Res 143:263–279, 1978.

83. Kawamura Y, Yamamoto T: Studies on neural mechanisms of the gustatory-salivary reflex in rabbits. J Physiol 285:35–47, 1978.

84. Mattes RD: Sensory influences on food intake and utilization in humans. Hum Nutr Appl Nutr 41A:77–95, 1987.

85. Harada S, Smith DV: Gustatory sensitivities of the hamster's soft palate. Chem Senses 17:37–51, 1992.

86. Smith DV, Hanamori T: Organization of gustatory sensitivities in hamster superior laryngeal nerve fibers. J Neurophysiol 65:1098–1114, 1991.

87. Brining SK, Belecky TL, Smith DV: Taste reactivity in the hamster. Physiol Behav 49:1265–1272, 1991.

88. Travers JB, Grill HJ, Norgren R: The effects of glossopharyngeal and chorda tympani nerve cuts on the ingestion and rejection of sapid stimuli: An electromyographic analysis in the rat. Behav Brain Res 25:233–246, 1987.

89. Berridge KC, Grill HJ: Isohedonic tastes support a two-dimensional hypothesis of palatability. Appetite 5:221–231, 1984.

90. Boggs DF, Bartlett D Jr: Chemical specificity of a laryngeal apneic reflex in puppies. J Appl Physiol 53:455–462, 1982.

91. Storey AT, Johnson P: Laryngeal water receptors initiating apnea in the lamb. Exp Neurol 47:42–55, 1975.

92. Shingai T, Miyaoka Y, Shimada K: Diuresis mediated by the superior laryngeal nerve in rats. Physiol Behav 44:431–433, 1988.

93. Smith DV, McBurney DH: Gustatory cross-adaptation: Does a single mechanism code the salty taste? J Exp Psychol 80:101–105, 1969.

94. Smith DV, Bealer SL: Sensitivity of the rat gustatory system to the rate of stimulus onset. Physiol Behav 15:303–314, 1975.

95. Halpern BP: Tasting and smelling as active, exploratory sensory processes. Am J Otolaryngol 4:246–249, 1983.

96. Smith DV, Shipley MT: The gustatory system. In: Conn PM (ed): Neucoscience in Medicine. Philadelphia: JB Lippincott, 1995.

97. Smith DV, Shipley MT: Anatomy and physiology of taste and smell. J Head Trauma Rehabil 7:1–14, 1992.

98. Smith DV, Travers JB, Van Buskirk RL: Brainstem correlates of gustatory similarity in the hamster. Brain Res Bull 4:359–372, 1979.

11

Taste Testing in Clinical Practice

JANNEANE F. GENT, Ph.D.
MARION E. FRANK, Ph.D.
APRIL E. MOTT, M.D.

As discussed earlier in this text, a number of university-based taste and smell clinics have been developed, where patients can receive comprehensive evaluations of chemosensory function. The evaluation of taste symptoms that patients receive at these clinics differs in many ways from the evaluation described in the medical literature. In general, the taste problems that emerge from case reports and case series are *secondary* to some other medical condition, ie, they are part of an unevaluated symptom cluster and not the focus of any objective testing. In an effort to catalog the nature of taste problems currently reported, we examined 33 articles[1] from the past decade that are case reports or case series of patients with taste dysfunction related to therapeutic intervention, disorders of the facial nerves, or disorders of the oral cavity. As can be seen in Table 11–1, the taste problem in over half of these is of a reduced ability to taste (hypogeusia)[2-16] or a complete absence of taste sensation (ageusia) on some part of the tongue or palate.[17-20] In one third of the articles, the reported taste dysfunction is a dysgeusia or "altered taste."[12,21-30] Of the remaining articles, only 2 report a "hypersensitivity"[31,32]; one complaint is ambiguously reported as a "change,"[33] and one reports other sensations (eg, burning mouth syndrome[34]).

In those case reports where the taste symptom did receive objective scrutiny, it was usually in the form of one test: a threshold test using either chemical stimuli or electrogustometry (eg, see Table 11–1). Of the 20 articles in Table 11–1 where objective tests of taste acuity were performed, 16 presented data from threshold testing. In 8 of these reports, the method used to obtain thresholds was electrogustometry. The other 8 reports of threshold testing described detection and/or recognition thresholds for chemical stimuli, with 4 of these employing some form of Henkin's three-drop technique.[35] In this sample of articles, the most popular form of taste function evaluation was patient self-report (Table 11–1). This was the case for 8 of 11 of the articles containing patient complaints of dysgeusia. Similarly, no tests of taste sensitivity were included in 4 out of 18 articles where patients described

Table 11–1 A Sample of Taste Problems and Evaluation Methods

| TASTE DYSFUNCTION | EVALUATION METHOD | | | | |
| | THRESHOLD | | DIRECT SCALING | SELF-REPORT | UNKNOWN |
	CHEMICAL	ELECTRO-GUSTOMETRIC			
Hypogeusia	5[2–6]	4[7–10]	2[4,11]	4[12–15]	1[16]
Ageusia	1[17]	1[18]		1[19]	1[20]
Dysgeusia		3[21–23]		8[12,24–30]	
Hypergeusia	1[31]		1[32]		
"Change"	1[33]				
Other			1[34]		

Entries are number of citations for that cell, eg, three papers[21–23] report "dysgeusia" evaluated using "electrogustometry." Note that some articles report more than one symptom (eg, "dysgeusia" and "hypogeusia"[12]) and use more than one method of symptom evaluation (eg, "chemical threshold" and "direct scaling"[4]). These papers appear in the table twice.

a reduced or absent taste sensitivity. Only 4 of the 33 articles included an assessment of above-threshold sensitivity.

Clinics specializing in taste and smell problems have developed tests that are being used to evaluate hundreds of patients yearly. We review here the taste evaluation procedures of four chemosensory clinics (see Table 11–2), chosen because they have published their methods, diagnostic criteria, and normative data. We describe in some detail and provide references for psychophysical techniques used to measure taste thresholds (both to chemical and electrogustometric stimuli) and the intensity of above-threshold taste sensations. We discuss the use of directed questioning to help clarify patient-reported dysgeusia, a symptom for which no objective measures exist.

TASTE INTENSITY PROBLEMS (HYPOGEUSIA, HYPERGEUSIA, AGEUSIA)

Clarification of the Patient-Reported Symptom

It has been the collective experience of the taste and smell clinics that the majority of patients who present with gustatory complaints are not suffering from a loss of taste function that can be measured.[36,37] For example, two thirds of the patients seen at the University of Pennsylvania (UPenn) Smell and Taste Clinic presented with a complaint that included some loss of taste function. However, after chemosensory testing, fewer than 4% of all of the patients had a measurable gustatory deficit, whereas 71% had a measurable olfactory deficit.[36]

In many cases, the presenting complaint of "I can't taste" actually reflects the common use of "taste" to mean "flavor," which includes not only how salty, sour,

Table 11–2　Clinic Evaluation of Regional Taste Function

TEST METHODS	CCRC[41]	UPENN[36,43]		MONELL[49]	NIHOUNU[39]	
Delivery	Cotton swab	Pipette	Electrode	Cotton swab	Filter paper	Electrode
Quality						
Salty	NaCl (1.0)	NaCl (0.512)		NaCl (1.0)	NaCl (0.05–3.4)	
Sweet	Sucrose (1.0)	Sucrose (1.28)		Sucrose (1.0)	Sucrose (0.009–2.3)	
Sour	Citric acid (0.03)	Citric acid (0.041)		Citric acid (0.018)	Tartaric acid (0.001–0.53)	
Bitter	Quinine HCl (0.001)	Caffeine (0.041)		$Quinine_2$ SO_4 (0.00018)	Quinine HCl (0.000025–0.1)	
Locations						
Tongue						
Tip	Yes	Yes	Yes (20)	Yes	Yes	Yes (20)
Edge	Yes	No	No	Yes	No	No
Posterior	No	Yes	No	No	Yes	Yes (40)
Palate	Yes	No	No	Yes	Yes	Yes (100)
Measure	Quality ID & intensity rating	Quality ID	Detection threshold	Quality ID & intensity rating	Quality ID & detection threshold	Detection threshold

CCRC = Connecticut Chemosensory Clinical Research Center, Farmington; UPenn = University of Pennsylvania Smell and Taste Center, Philadelphia; Monell = Monell Chemical Senses Center, Philadelphia; NihonU = Department of Otorhinolaryngology. Nihon University School of Medicine, Tokyo.
*Molarity of chemical and maximum microamperes of electrical stimuli in parentheses.

sweet, or bitter something is, but other oral-cavity sensations including odor, texture, and pleasantness. Most patients, on being questioned, report that they can in fact "taste" salty, sour, sweet, and bitter. An interview is, therefore, an important part of most clinics' initial evaluation, as it helps clarify and define the symptom that the patient is reporting.

Ideally, the interview should be able to "screen out" those patients who are unlikely to have taste perception problems. Gent, Goodspeed, Zagraniski, and Catalanotto[38] examined how well screening questions predicted actual taste function as measured by clinical testing. Results indicate that "Do you have trouble tasting?" is a poor screening question. For example, the probability that a patient has a gustatory deficit given a positive response to this question (positive predictive value) was found to be less than 10%.[38] For the reason indicated above, ie, confusion of "taste" and "flavor," patients tend to overreport "taste" problems. A better screening question for quantitative taste function is "Do you have trouble tasting salt, sweet, sour, bitter?" When the self-report for each quality was pooled, the negative predictive value of this question was found to be 94%,[38] ie, a patient responding "No" to this question has a 94% chance of not having any measurable taste deficit for sweet, salty, sour, and bitter. A positive response to this question suggests that further testing of taste function is necessary.

Regional Testing of Taste Function

Several thousand taste buds are distributed in specific receptor fields throughout the oral cavity. Innervation of the gustatory system involves three cranial nerves bilaterally. The chorda-tympani branch of the facial nerve innervates the fungiform papillae along the anterior two thirds of the tongue and the greater-superficial-petrosal branch of the facial nerve innervates taste buds on the palate. The glossopharyngeal nerve innervates the foliate papillae along the posterior edge of the tongue and the large circumvallate papillae across the posterior tongue. The vagus nerve innervates taste buds in the throat.

Regional (spatial) testing of taste function on some or all of the lingual and palatal receptor fields is part of the test battery for the four clinics surveyed (see Table 11–2). Because all receptor fields are capable of responding to stimuli representing basic taste qualities (eg, salty, sweet, sour, and bitter), although not necessarily with equal sensitivity,[39,40] the four clinics use representative chemical stimuli (eg, NaCl for salty, sucrose for sweet, an organic acid for sour, and a quinine salt or caffeine for bitter) to test function in each area. UPenn and Nihon University (NihonU) clinics also test for absence or presence of regional taste response using electrogustometric stimuli (see Table 11–2).

Chemical Stimuli

The stimulus delivery method used at the Connecticut Chemosensory Clinical Research Center (CCCRC)[41] is also used at the Monell Chemical Senses Center (Monell). A sterile cotton swab is dipped in one of four solutions, then painted on one of six lingual or palatal locations. Each of four tongue quadrants (anterior and posterior, left and right) and two palatal areas (left and right of midline) are stimulated. Concentrations used are well above threshold (see above and Table 11–2). The patient is asked to identify and then rate the intensity of each solution on a 10-point (CCCRC) or 13-point (Monell) scale from "no taste" (0) to extremely strong (9 [CCCRC] and 13 [Monell]). This test is used as a screening test to show presence or absence of function in areas predominantly innervated by a different cranial nerve. It is also used to indicate left-right differences in function. At the CCCRC clinic, results of this spatial test are compared to a control population.

There appears to be some overlap in the innervation fields of the chorda tympani and glossopharyngeal in the area of the foliate papillae.[39] The anterior foliate area may be exclusively innervated by the chorda tympani, whereas the posterior foliate papillae are innervated exclusively by the glossopharyngeal nerve.[42] For this reason, the UPenn regional test avoids the foliate papillae. Tests of the anterior fungiform papilla and one circumvallate papilla bilaterally assess chorda tympani and glossopharyngeal function, respectively[36] (Table 11–2). The palate is not tested. A 20-mL pipette is used to deliver stimuli to each of the lingual locations. As is the case at CCCRC and Monell, the stimuli for the spatial test are the strongest concentrations used in the whole-mouth intensity rating test[36,43] (see Table 11–3). The patient's task is to correctly identify each stimulus after it is placed on one of the four tongue areas. The number of correct identifications is compared only within nerve fields (ie, between left and right for the chorda tympani or glosso-pharyngeal nerve fields). Significant differences in number correct indicate side-to-side differences.

At CCCRC, Monell, and UPenn, regional testing is one of several used to provide an overview of the patient's gustatory function. At NihonU,[39] all taste testing is regional. The stimulus is delivered with a solution-soaked disc of filter paper, which controls solution spread. Five concentrations of each of four tastants

Table 11–3   Clinic Evaluation of Whole-Mouth Taste Function: Detection Threshold

STIMULUS QUALITY	CLINIC	
	UPENN[43]	MONELL[49]
Salty	NaCl (0.000125–0.512)	NaCl (0.00001–1.0)
Sweet	Sucrose (0.000312–1.28)	Sucrose (0.00001–1.0)
Sour	Citric acid (1.56×10^{-7}–0.0412)	Citric acid (0.000001–0.018)
Bitter	Not tested	Quinine$_2$ SO$_4$ (5.6×10^{-9}–0.00018)

UPenn = University of Pennsylvania Smell and Taste Center, Philadelphia; Monell = Monell Chemical Senses Center, Philadelphia. Molarity of chemical stimuli in parentheses.

are used (Table 11–2). Solutions are presented from weak to strong on each tongue and soft-palate location. The patient tastes the solution and then points to a chart containing the following response choices: sweet, salty, sour, bitter, undefined taste, and no taste. A patient's taste function is classified as "hypergeusia"; "normal"; or "low-," "medium-," or "high-hypogeusia" based on the solution level for each quality at which the patient detects a taste (the detection threshold) compared to norms established by the clinic.[39]

The taste system exhibits spatial summation for small areas of stimulation[44]; the more papillae a stimulus contacts, the stronger it tastes. For a regional test of taste function, the advantage of the filter paper method over pipette or swab is in the control over stimulus amount and placement. This control is crucial if one side of the tongue is to be compared to the opposite side. For such a comparison to be meaningful, bilateral symmetry in stimulus placement is crucial.

Electrogustometry

UPenn and NihonU have used electrogustometry extensively as part of their taste test batteries. Most people detect a metallic, salty, or sour taste when regions containing taste buds are stimulated with the anode. The pros and cons of electrogustometric evaluation have been reviewed recently.[45] Both UPenn and NihonU use this method to measure detection thresholds, although the UPenn test is restricted to the two sides of the anterior tongue.

Whole-Mouth Tests of Taste Function

Taste complaints occur when ordinary eating and drinking produce abnormal oral sensations. A patient describing hypogeusia, for example, may or may not be aware of any regional deficits in taste. In fact, it has been observed that patients who have one (or more) nonfunctioning receptor fields very often do not report a reduced sense of taste.[39,46] It has been hypothesized that "cross-inhibition" in the taste system could account for this phenomenon.[47] In other words, the chorda tympani and glossopharyngeal nerves exert reciprocal inhibitory influences, and if one receptor area is no longer able to respond to stimuli, it is also no longer inhibiting the responses of neighboring receptors. As a result, the net sensation experienced is similar to that experienced when all receptor fields were operating.

Although a regional test is the best source of information about gustatory function in the various taste receptor fields, a whole-mouth test is a way to quantify a patient's "real-world" taste function. CCCRC, Monell, and UPenn use whole-mouth tests of taste function, either a threshold or a scaling procedure, or both (Tables 11–2 and 11–3).

Threshold Tests

It is theoretically possible to measure two different threshold responses for any particular taste stimulus: a "detection" threshold (the concentration at which the stimulus is first "tasted") and a "recognition" threshold (the concentration at which the stimulus is recognized as having a particular taste, eg, salty, sour, sweet, or bitter). The major problem with recognition-threshold procedures is that they are subject to response bias.[48] For example, a subject can easily bias the results of a four-alternative, forced-choice recognition task by always guessing "bitter" when the taste is "uncertain." This violates the underlying assumption that the subject's guesses are randomly distributed among the four taste qualities and will artificially reduce the recognition threshold of "bitter."

A two-alternative, forced-choice detection threshold procedure, the method preferred by Monell and UPenn, requires a taste/no taste decision. Briefly, solutions are sampled from two cups, one containing a tastant and the other water, with water rinses preceding each sample. The patient is "forced" to choose the one with a "taste," even if it is a guess. When a choice is correct, solution concentration is decreased, but if the choice is incorrect the concentration is increased on the next trial. The detection threshold is "tracked" in this way until some criterion number of reversals is reached (eg, Cowart[49]). The Monell clinic determines a detection threshold for NaCl, sucrose, citric acid, and quinine sulfate; UPenn excludes a bitter compound. Each follows a standard psychophysical paradigm for a response that is relatively free of bias (two-alternative forced choice) and adaptation (water rinse after every stimulus).

None of the clinics surveyed here use Henkin's three-drop technique[35] or the Harris-Kalmus threshold procedure,[50] both of which are popular in the medical literature. Each of these has procedural problems with response bias and/or stimulus adaptation that tend to bias the results.[48]

Suprathreshold Tests

It is possible to extract information about the sensitivity of the taste system to small differences in stimuli that are above threshold. Sensory experience can be scaled directly by using patients as their own perceived intensity-measuring devices (for a general discussion of psychophysical scaling methods, see Engen,[51] Marks,[52] or McBurney[53]).

Direct scaling tasks can take different forms, as can be seen in the suprathreshold procedures used by the clinics (Table 11–4). One form asks the patient to categorize a sensory experience with numbers associated with adjectives such as "weak," "moderate," or "strong." Monell and UPenn ask patients to rate the intensity of three (Monell[49]) or five (UPenn[36]) concentration levels of basic taste stimuli on a 13-point (Monell) or 9-point (UPenn) scale. The strongest solutions used in this test are also used as the stimuli in each clinic's regional taste test (see above and Table 11–2).

Category scaling will indicate which stimuli taste strong and which taste weak, but not how much stronger one stimulus is compared to another. Category scales are not ratio scales, ie, a stimulus given a rating of 8 is not necessarily twice as strong as a stimulus rated 4. This is a problem if the information from the scaling task is needed to calculate a dose-response (psychophysical) function for the gustatory system. Given proper instructions, patients can produce estimates of taste intensity magnitude that are consistent with a ratio scale.[54] For example, a patient may be asked to assign a number to the maximum intensity of each solution tasted, using numbers in such a way that if a solution tastes twice as strong, it should be given

Table 11–4   Clinic Evaluation of Whole-Mouth Taste Function: Suprathreshold Test

	CLINIC		
	CCCRC[41]	UPENN[36,43]	MONELL[49]
Stimulus quality			
Salty	5 NaCl (0.01–1.0)	5 NaCl (0.032–0.512)	3 NaCl (0.01–1.0)
Sweet	5 sucrose (0.01–1.0)	5 sucrose (0.08–1.28)	3 sucrose (0.01–1.0)
Sour	5 citric acid (0.00032–0.032)	5 citric acid (0.00256–0.041)	3 citric acid (0.0018–0.018)
Bitter	6 quinine HCl (0.0000032–0.001) 4 PROP (0.000056–0.00018)	5 caffeine (0.00256–0.041)	3 quinine$_2$ SO$_4$ (0.000018–0.00018)
Response measure			
	Magnitude match to 1000-Hz tones (38 dB to 98 dB)	Rating on 9-point category scale	Rating on 13-point category scale

CCCRC = Connecticut Chemosensory Clinical Research Center, Farmington; UPenn = University of Pennsylvania Smell and Taste Center, Philadelphia; Monell = Monell Chemical Senses Center, Philadelphia. Table entries for test stimuli include number of concentration levels per compound (eg, 5 NaCl) and molar concentration range in parentheses.

twice as large a number. Their task is to make the numbers proportional to the taste intensity (see Stevens[54]).

Direct scaling tests reveal the ability to discriminate among stimulus intensity levels and whether stimuli of one quality (eg, sweet) are stronger or weaker than another (eg, salty). Direct scaling cannot determine whether all taste stimuli are being perceived at a level that is abnormal, but magnitude matching can.[55] In magnitude matching, a patient rates the intensity of taste stimuli and stimuli from another sensory modality, such as the loudness of tones on the same scale. The CCCRC clinic asks patients to rate the strength of NaCl, sucrose, citric acid, quinine HCl, and 6-*n*-propylthiouracil solutions and the loudness of 1000-Hz tones (Table 11–4). Assuming a normal sense of hearing, the results of this cross-modality matching test reveal how strong taste stimuli are relative to the loudness of the auditory stimuli.

One concern with numerical magnitude estimation tasks that are not tied to adjectives is a patient's idiosyncratic use of numbers. In cases where this is a concern, patients can be asked to match taste intensity to line length.[49,56] A general concern with magnitude matching is its difficulty for untrained subjects.[57,58] Rating scales (category scales, line scales, and magnitude estimation) were found to be equally sensitive in measuring intensity differences[57–59]; however, subjects found magnitude estimation to be "hardest to understand," "slowest" to use to complete the rating task, and "most restrictive."[58]

The taste system signals quality as well as intensity, and all clinics include a quality identification test (ie, the patient is asked to identify each stimulus as "salty," "sweet," "sour," "bitter," or "other") as part of the direct scaling tests and/or the test of regional function (Table 11–2).

TASTE QUALITY DISTORTION (DYSGEUSIA)

The first step in evaluating a dysgeusia is to establish that the quality distortion is not olfactory in origin. For example, a patient complaint of a persistently sweet or bitter taste is more likely to be gustatory than if the patient complains of a foul or smoky taste.

Clinics question patients about the history, onset, and duration of a dysgeusia. At UPenn, dysgeusias are classified as: (1) "transient or nontransient," (2) "stimulated or nonstimulated" (ie, induced by oral stimulation, or not), and (3) "identifiable or not identifiable" (ie, describable as "salty, sour, sweet, bitter, and/or metallic," or not[36]).

DIAGNOSIS OF TASTE DISORDERS

Although the diagnostic weight attached to particular parts of the evaluation varies, each part of an evaluation of gustatory function contributes to the diagnosis of a patient's taste function.

The NihonU clinic relies on regional testing to diagnose a patient's taste function. At this clinic, a diagnosis based on five detection thresholds is made for each of three lingual and palatal locations bilaterally (Table 11–2). The criteria used in making a diagnosis have been published along with the normative data on which they are based.[39] Normative data used for diagnosis are not age-corrected, although age effects have been observed.

Taste evaluations at CCCRC, Monell, and UPenn focus on whole-mouth taste function. Detection thresholds are given the greatest weight in the diagnosis of taste function at Monell (Table 11–3). Thresholds are considered abnormal if they are within the upper 2.5 percent of threshold distributions of normal subjects.[49] Sex-corrected norms are not used, although normal thresholds for males and females differ for citric acid. Taste intensity ratings for individual stimuli are given the most diagnostic weight at UPenn (Table 11–3). Taste function is considered abnormal if the average rating for at least two concentrations is at or below the 2.5th percentile for controls. Ratings of taste stimuli on a 9-point hedonic scale from "like extremely" to "dislike extremely" and quality identification serve as confirmatory tests. The hedonic rating has proven to be useful in cases where a patient is unable or unwilling to rate the intensity of a stimulus (R.G. Settle, personal communication, May 26, 1992). "Total taste" intensity is given the most diagnostic weight at CCCRC. Data are normalized to loudness ratings as described fully in Marks et al,[55] and "total taste" is the summed magnitude estimates for concentration series of

NaCl, sucrose, citric acid, and quinine HCl. Patients performing above the 15th percentile are considered to have normal gustatory function.

CONCLUSION

A comprehensive evaluation of gustatory dysfunction begins with clarification and definition of the reported taste symptom, using a self-administered questionnaire and/or an interview. Evaluation continues with assessment of regional and/or whole-mouth taste function.

The four clinics discussed have chosen test compounds that are still considered by most of the chemosensory community to be the best exemplars of basic taste qualities: salty, sour, sweet, and bitter. This is in spite of years of discussion over the true nature of taste qualities.[60,61] The concentration ranges chosen for compounds used in common are overlapping. For example, in Table 11–4, the strongest solutions used in the whole-mouth tests are either identical or within one order of magnitude of molarity. Maximum concentrations approximate the intensity saturation point (most intense response possible) of the taste system.

Directed questioning and threshold and suprathreshold tests are considered necessary and sufficient for a comprehensive assessment of taste function. However, although diagnostic tests of taste function used in specialized clinics (see appendix) employ state-of-the-art psychophysical methods, data on the validity, reliability, sensitivity, and specificity of the tests is sparse.[62] In addition, taste evaluations in chemosensory clinics are more comprehensive, but more cumbersome and time-consuming than taste testing reported in case studies, and are not widely used outside of these special clinics. Practical instruments for taste testing incorporating essential aspects of specialized comprehensive evaluations are not yet available for transfer to general clinical practice.

REFERENCES

1. Mott AE, Grushka M, Sessle BJ: Diagnosis and management of taste disorders and burning mouth syndrome. Dent Clin North Am 37:33–71, 1993.
2. Abu-Hamdan DK, Desai H, Sondheimer J, et al: Taste acuity and zinc metabolism in captopril-treated hypertensive male patients. Am J Hypertens 1:303S–308S, 1988.
3. Fujimura A, Kajiyama H, Tateishi T, Ebihara A: Circadian rhythm in recognition threshold of salt taste in healthy subjects. Am J Physiol 259:R931–R935, 1990.
4. Mela DJ: Gustatory function and dietary habits in users and nonusers of smokeless tobacco. Am J Clin Nutr 49:482–489, 1989.
5. Peretianu D, Deleanu A, Tanase F: Gustative sensitivity to glucose improvement in diabetics after sulphonylurea. Rev Roum Physiol 27:115–120, 1990.
6. Wayler AH, Perlmuter LC, Cardello AV, et al: Effects of age and removable artificial dentition on taste. Spec Care Dent 10:107–113, 1990.
7. Grant R, Miller S, Simpson D, et al: The effect of chorda tympani section on ipsilateral and contralateral salivary secretion and taste in man. J Neurol Neurosurg Psychiatry 52:1058–1062, 1989.
8. Imamine T, Okuno M, Moriwaki H, et al: Plasma retinol transport system and taste acuity in patients with obstructive jaundice. Gastroenterol Jpn 25:206–211, 1990.
9. Ralli G, Magliulo G, Persichetti S, et al: Electrogustometry in hemodialysis patients. Acta Otorhinolaryngol Belg 39:822–831, 1985.
10. Soni NK, Chatterji P: Gustotoxicity of bleomycin. *ORL* 47:101–104, 1985.
11. Lang NP, Catalanotto FA, Knopfli RU, Antczak AAA: Quality-specific taste impairment following the application of chlorhexidine digluconate mouth rinses. J Clin Periodontol 15:43–48, 1988.

12. Berman JL: Dysosmia, dysgeusia and diltiazem. Ann Intern Med 102:717, 1985.
13. Axelrod FB, Pearson J: Congenital sensory neuropathies. AJDC 138:947–954, 1984.
14. Hepso HU, Bjornland T, Skoglund LA: Side-effects and patient acceptance of 0.2% versus 0.1% chlorhexidine used as postoperative prophylactic mouthwash. Int J Oral Maxillofac Surg 17:17–20, 1988.
15. Strong MJ, Noseworthy JH: Hemiageusia, hemianaesthesia and hemiatrophy of the tongue. Can J Neurol Sci 13:109–110, 1986.
16. Zumkley H, Vetter H, Mandelkow T, Spieker C: Taste sensitivity for sodium chloride in hypotensive, normotensive and hypertensive subjects. Nephron 47(suppl 1):132–134, 1987.
17. Kassirer MR, Such RvP: Persistent high-altitude headache and aguesia [sic] without anosmia. Arch Neurol 46:340–341, 1989.
18. Kikuchi T, Kusakari J, Kawase T, Takasaka T: Electrogustometry of the soft palate as a topographic diagnostic method for facial paralysis. Acta Otolaryngol 458(suppl):134–138, 1988.
19. Ewing RC, Janda SM, Henann NE: Ageusia associated with transdermal nitroglycerin. Nature 196:74–75, 1989.
20. Duhra P, Foulds IS: Methotrexate-induced impairment of taste acuity. Clin Exp Dermatol 13:126–127, 1988.
21. Axell T, Nilner K, Nilsson B: Clinical evaluation of patients referred with symptoms related to oral galvanism. Swed Dent J 7:169–178, 1983.
22. Nilner K, Nilsson B: Intraoral currents and taste thresholds. Swed Dent J 6:105–113, 1982.
23. Rieder C: Eine seltene Komplikation: Geschmacks-storung nach Tonsillektomie. Laryngol Rhinol 60:342, 1981.
24. El-Deiry A, McCabe BF: Temporal lobe tumor manifested by localized dysgeusia. Ann Otol Rhinol Laryngol 99:586–587, 1990.
25. Frankel DH, Mostofi RS, Lorincz AL: Oral Crohn's disease: Report of two cases in brothers with metallic dysgeusia and a review of the literature. J Am Acad Dermatol 12:260–268, 1985.
26. Gelenberg AJ, Kane JM, Keller MB, et al: Comparison of standard and low serum levels of lithium for maintenance treatment of bipolar disorder. N Engl J Med 321:1489–1493, 1989.
27. Goy JJ, Finci L, Sigwart U: Dysgeusia after high dose dipyridamole treatment. Drug Res 35:854, 1985.
28. Levenson JL, Kennedy K: Dysosmia, dysgeusia and nifedipine. Ann Intern Med 102:135–136, 1985.
29. Miller SM, Naylor GJ: Unpleasant taste—a neglected symptom in depression. J Affect Dis 17:291–293, 1989.
30. Schon F: Involvement of smell and taste in giant cell arteritis. J Neurol Neurosurg Psychiatry 51:1594, 1988.
31. Ovenson L, Hannibal J, Sorensen M: Taste thresholds in patients with small-cell lung cancer. J Cancer Res Clin Oncol 117:70–72, 1991.
32. Mattes RD, Christensen CM, Engelman K: Effects of hydrochlorothiazide and amiloride on salt taste and excretion (intake). Am J Hypertens 3:436–443, 1990.
33. Heise E, Schnuch A: Taste and olfactory disturbances after treatment for acne with isotretinoin, a 13-cis-isomer of retinoic acid. Eur Arch Otorhinolaryngol 247:382–383, 1990.
34. Grushka M, Sessle BJ: Taste impairment in burning mouth syndrome. Gerodontology 4:256–258, 1988.
35. Henkin RI, Gill JR Jr, Bartter FC: Studies on taste thresholds in normal man and in patients with adrenal cortical insufficiency: The role of adrenal cortical steroids and serum sodium concentration. J Clin Invest 42:727–735, 1963.
36. Deems DD, Doty RL, Settle G, et al: Smell and taste disorders, a study of 750 patients from the University of Pennsylvania Smell and Taste Center. Arch Otolaryngol Head Neck Surg 117:519–528, 1991.
37. Goodspeed RB, Gent JF, Catalanotto FA: Chemosensory dysfunction: Clinical evaluation results from a taste and smell clinic. Postgrad Med 81:251–260, 1987.
38. Gent JF, Goodspeed RB, Zagraniski RT, Catalanotto FA: Taste and smell problems: Validation of questions for the clinical history. Yale J Biol Med 60:27–35, 1987.
39. Tomita H, Ikeda M, Okuda Y: Basis and practice of clinical taste examinations. Auris Nasus Larynx (Tokyo) 13(suppl I):S1–S15, 1986.
40. Gent JF, Bartoshuk LM: Sweetness of sucrose, neohesperiden dihydrochalcone and saccharin is related to genetic ability to taste the bitter substance 6-n-propylthiouracil. Chem Senses 7:265–272, 1983.
41. Bartoshuk LM: Clinical evaluation of the sense of taste. Ear Nose Throat J 68:331–337, 1989.
42. Catalanotto FA, Lecadre Y, Devonshire F, Bartoshuk L: Foliate papillae taste perception in humans. Chem Senses 16:508, 1991.

43. Settle RG, Quinn MR, Brand JG, et al: Gustatory evaluation of cancer patients: Preliminary results. In: van Eys J, Seelig MS, Nichols BL (eds): Nutrition and Cancer. New York: Spectrum Publications, 1979.
44. Smith DV: Taste intensity as a fuction of area and concentration. J Exp Psychol 87:163–171, 1971.
45. Frank ME, Smith DV: Electrogustometry: A simple way to test taste. In: Getchell TV, Doty RL, Bartoshuk LM, Snow JB Jr (eds): Smell and Taste in Health and Disease. New York, Raven Press, 1991.
46. Ostrom KM, Catalanotto FA, Gent JF, et al: Effects of oral sensory field loss on taste scaling ability. Chem Senses 10:459, 1985.
47. Miller IJ Jr, Bartoshuk LM: Taste perception, taste bud distribution, and spatial relationships. In: Getchell TV, Doty RL, Bartoshuk LM, Snow JB Jr (eds): Smell and Taste in Health and Disease. New York, Raven Press, 1991.
48. Weiffenbach JM, Wolf RO, Benheim AE, Folio CJ: Taste threshold assessment: A note on quality specific differences between methods. Chem Senses 8:151–159, 1983.
49. Cowart BJ: Relationships between taste and smell across the adult life span. Ann NY Acad Sci 561:39–55, 1989.
50. Harris H, Kalmus H: The measurement of taste sensitivity to phenylthiourea (PTC). Ann Eugen 15:24–31, 1949.
51. Engen T: Psychophysics II: Scaling methods. In: Kling JW, Riggs LA (eds): Woodworth and Schlosberg's Experimental Psychology. New York: Holt, Rhinehart and Winston, 1971.
52. Marks LE: Sensory Processes: The New Psychophysics. New York: Academic Press, 1974.
53. McBurney DH: Experimental Psychology. Belmont, CA: Wadsworth Publishing Co, 1983.
54. Stevens SS: Sensory scales of taste intensity. Percept Psychophys 6:302–308, 1969.
55. Marks LE, Stevens JC, Bartoshuk LM, et al: Magnitude-matching: The measurement of taste and smell. Chem Senses 13:63–87, 1988.
56. Weiffenbach JM, Cowart BJ, Baum BJ: Taste intensity perception in aging. J Gerontol 41:460–468, 1986.
57. Lawless HT, Malone GJ: A comparison of rating scales: Sensitivity, replicates and relative measurement. J Sens Stud 1:155–174, 1986.
58. Lawless HT, Malone GJ: The disriminative [sic] efficiency of common scaling methods. J Sens Stud 1:85–98, 1986.
59. Lawless HT: Logarithmic transformation of magnitude estimation data and comparisons of scaling methods. J Sens Stud 4:75–86, 1989.
60. McBurney DH, Gent JF: On the nature of taste qualities. Psych Bull 86:151–167, 1979.
61. Schiffman SS, Erickson RP: A theoretical review: A psychophysical model for gustatory quality. Physiol Behav 7:617–633, 1971.
62. Doty RL: Diagnostic tests and assessment. J Head Traum Rehabil 7:47–65, 1992.

APPENDIX

Chemosensory clinics listed in the *1991 Association for Chemoreception Sciences Membership Directory* include:

Chemosensory Clinical Research Center
Monell Chemical Senses Center
3500 Market St.
Philadelphia, PA 19104-3308

Clinical Olfactory Research Center
SUNY Health Science Center at Syracuse
766 Irving Ave.
Syracuse, NY 13210

Connecticut Chemosensory Clinical Research Center
University of Connecticut Health Center
Farmington, CT 06032

MCV Taste and Smell Clinic
Medical College of Virginia
Richmond, VA 23298-0551

Nasal Dysfunction Clinic
University of California, San Diego
Medical Center
225 Dickinson St.
San Diego, CA 92103

National Institute of Dental Research
National Institutes of Health
NIH Building 10 Room 1N-114
Bethesda, MD 20892

Rocky Mountain Taste and Smell Center
University of Colorado Health Science
 Center
4200 East Ninth Ave.
Denver, CO 80262

University of Cincinnati Taste and Smell Center
University of Cincinnati College of Medicine
231 Bethesda Ave.
Cincinnati, OH 45267-0528

University of Pennsylvania Smell and Taste Center
Hospital of University of Pennsylvania
3400 Spruce St.
Philadelphia, PA 19104-4283

12

Burning Mouth Syndrome

MIRIAM GRUSHKA, M.Sc., D.D.S., Ph.D.
JOEL B. EPSTEIN, D.M.D., MS.D.

Burning mouth syndrome (BMS) is an intraoral pain disorder usually unaccompanied by mucosal lesions or other clinical signs. For the majority of BMS patients, the onset of the pain is spontaneous.[1] More than one oral site is usually affected, with the anterior two thirds of the tongue, the anterior hard palate, and the mucosa of the lower lip most frequently involved.[2-6] The distribution of oral sites does not appear to affect the course of the disorder[7,8] or the response to treatment.[7] Other symptoms frequently accompany the oral burning and include dry mouth, thirst, and altered or dysguesic taste.[1,9,10] For many BMS patients, the burning begins by late morning and usually reaches maximum intensity by evening.[1,4,9,11-14] The burning, which is usually continuous once it has started for the day, often makes falling asleep at night difficult for many patients.[10,15] Self-reported mood changes that include irritability, depression, alterations in eating habits, and a decreased desire to socialize occur more frequently in BMS patients than in age- and sex-matched control subjects, and may be related to altered sleep patterns.[1,4,16]

Pain levels in BMS have been found to be intense[3,4,10,17,18] and, in one study, similar in intensity, although of different quality, to toothache pain.[3] This latter study also demonstrated that when BMS patients were compared with asymptomatic control subjects, the BMS patients showed changes in certain personality characteristics similar to those observed in other groups of chronic pain patients; moreover, these characteristics were found to be exacerbated by increased levels of pain. In contrast, no significant correlation has been found between pain intensity (or intensity of other oral sensations such as taste disturbances, see below) and a subject's previous history of psychological/psychiatric treatment.[10]

ALTERATIONS IN TASTE

Many studies have documented that BMS patients frequently complaint of a persistent or altered (dysguesic) taste and/or alterations in the perception of sweet, sour, salty, and bitter.[1,4,9,10,12,13,19] In one study,[1] for example, it has been demonstrated that significantly more BMS patients (69%) than age and sex-matched control subjects (11%) reported either an alteration in taste perception or a persistently altered taste (dysqeusia). Of the BMS subjects with a taste complaint, the most common was either a persistent taste only (29%), an alteration in taste perception only (8%), or both types of changes (33%). The persistent taste was most frequently bitter, metallic, or a combination of both and was reduced in most BMS patients by rinsing or eating.

MENOPAUSE

Appromiately 90% of women who participate in experimental studies of BMS[4-6,11,16,18,20] are postmenopausal, with the greatest frequency of onset reported to be from 3 years before to 12 years following menopause.[1] One study[1] of postmenopausal women with BMS demonstrated that these women rated their menopausal symptoms as having been significantly more severe than a group of matched control subjects. However, in spite of these findings, no significant difference has been found between BMS and appropriate control subjects in any of the following factors: the number of years since menopause, the occurrence of surgical menopause, the usage of estrogen replacement therapy (ERT), the number of years of treatment by ERT, and the number of years passed since completion of ERT treatment.

DEMOGRAPHICS

Oral burning appears to be most prevalent in postmenopausal females, a conclusion based on the make-up of most samples in experimental studies[1,4,9,11,14,17–19,21,22]; the occurrence of oral discomfort, including burning, in 10% to 40%[21,23] or more[24] of women who attend centers for the treatment of menopausal symptoms; and the report of mouth burning by approximately 16% of female respondents aged 40 to 49 years in a general dental survey.[21] This figure is in contrast to the much lower overall prevalence of mouth burning (from 0.7% to 2.6%) that has been reported in epidemiological studies including a group of normal dental patients,[21] a healthy group of menopausal women,[25] a random sample of 1000 Canadians living in a large urban setting,[26] and in a recent large demographic study of orofacial pain in the United States (see Klausner[27] for review). Clearly, more study is necessary to determine the actual prevalence rate and gender ratio in the general population and to determine if postmenopausal females are overrepresented in experimental trials.

SPONTANEOUS REMISSION

Although there are only a few studies reporting spontaneous remission in BMS,[8,28] it does appear that at least a partial spontaneous remission of BMS may occur within 6 to 7 years after onset, in approximately one half[8] to two thirds[28] of subjects. In one study,[8] spontaneous remission was found to be preceded by a change in the pattern of burning from constant to episodic, with most subjects unable to identify a contributing factor for the change in pattern. No significant differences in age, sex, duration of disease, or distribution of burning sites were found in this study among individuals who experienced partial remission in their symptoms or recovered completely and those who continued to experience burning pain.

CLINICAL FINDINGS

Systemic Conditions

Although most studies[1,4,5,9,29] have reported no significantly higher prevalence in BMS of any specific medical condition, a higher occurrence of a wide variety of other health complaints, chronic pain conditions,[1,18,30–32] and medication usage[18,33] has been found. For example, one recent study[18] demonstrated a significant relationship in women between BMS and self-reported anemia, inadequate diet, chronic infection, and hormonal therapies, as well as mucosal ulcerative/erosive lesions and atrophy. In contrast, BMS in men in this study showed a significant relationship with central nervous system disturbances and gingivitis. In addition, significant associations were also found to be related to variables such as psychogenic factors, regurgitation, flatulence, and periodontitis.

Hematologic examinations of BMS subjects for nutritional deficiency or other systemic disorders[1,4,5,11,23,29,34] are frequently, although not always, (see, eg, Lamey and Lamb[4]) negative. It has, however, been found that more than 58% of BMS subjects demonstrate mildly abnormal results for immunologic features such as rheumatoid and antinuclear factor as well as complement levels[35] suggestive of a possible association between BMS and a connective tissue disorder[36] (see Ship et al[37] for review).

Local Factors

Although Grushka[1] has reported no significant differences on clinical examination between BMS and control subjects in any intraoral soft or hard tissues,[1] other studies[18,30,38] have reported a higher incidence of oral changes (eg, gingivitis; periodontitis; ulcerative/erosive lesions; geographic, fissured, scalloped, or erythematous tongue) in BMS subjects. This discrepancy may be the result of the subjective nature of the diagnosis of many of these soft tissue conditions and argues for the need for strict criteria for their diagnosis (eg, see Wolff et al[25]).

Salivary Features

It has long been suggested that dry mouth[18,39,40] may be a primary cause of BMS in older individuals. Although recent studies have found neither oral mucosal changes[25] nor decreased salivary flow to be associated with increasing age,[41-44] a decreased salivary flow rate has been found to be associated with a number of systemic diseases[45] and chronic medication usage.[45,46] This is especially marked in postmenopausal females,[41-44] the group most at risk for oral complaints including mouth burning.[47] On the other hand, most salivary flow rate studies in patients with BMS[4,48-51] have not demonstrated a significant decrease in salivary output. Many have, however, demonstrated (in stimulated and unstimulated whole saliva and stimulated parotid saliva) significant alterations in salivary levels of factors such as proteins, immunoglobulins, and phosphates as well as differences in saliva pH, buffering capacity, electrical resistance, and conductance when compared with control values.[51-55] Although the relationship of these alterations in salivary composition to BMS is unknown, it has been postulated that these alterations may be related to a selective rather than a gross change in salivary flow rate related to age or disease (eg, see Wu and Ship,[45] Navazesh et al,[56] and Närhi,[57]) and may offer a fruitful avenue for further research.

Taste Changes

As described earlier, many reports have referred to disturbances in the sense of taste in BMS subjects. Altered taste thresholds to electric stimuli have been reported previously[38,58] in subjects with oral complaints. More recently, Grushka et al[17] have demonstrated significant differences in threshold and suprathreshold levels of taste perception in BMS subjects compared with matched controls. At threshold concentrations the ability to taste sweet was significantly decreased (ie, threshold was elevated); at suprathreshold concentrations, both sweet and sour solutions were perceived to be more intense by the BMS subjects than the control subjects. Of note, these differences at suprathreshold concentrations were noted only between those BMS subjects who reported a taste dysgeusia and control subjects. Significant differences also were found between those BMS subjects with a dysgeusic taste complaint and those BMS subjects without such a complaint.

One likely reason to explain these findings, according to these investigators, is that approximately 60% of the BMS subjects in this study complained of a dysgeusic taste. At low solute concentrations, especially for sweet and sour, BMS subjects with dysgeusia may perceive their own dysgeusic taste in addition to the actual taste stimuli; at higher solute concentrations, the dysgeusic taste may become masked. This view was supported by further studies[17] that indicated that at low solute concentrations of sweet, bitter, and salt, BMS subjects with dysgeusia are more likely than those BMS subjects without dysgeusia to inappropriately identify the taste of the solution and to assign to it a higher intensity value. Whether successful medical management of BMS improves these taste disturbances may prove another fruitful avenue for future research.

Changes in Somatosensory Function

Burning pain, the chief reported symptom in BMS, is also the characteristic symptom of many posttraumatic nerve injuries. These latter conditions, however, may be associated with sensory abnormalities in features such as two-point discrimination, touch, temperature perception, pain threshold, and pain tolerance that may occur in addition to the burning pain.[59–61] In contrast to the above, no differences were found by Grushka et al[62] in any of these somatosensory modalities in BMS subjects when compared to control subjects at any of eight intraoral and facial sites tested, with the sole exception of heat pain tolerance, which was found to be significantly reduced at the tongue tip. Differences were found, however, in a more recent study using argon laser stimulation that revealed significant qualitative and quantitative changes in some sensory functions of BMS subjects compared to controls.[63] Studies such as these, which demonstrate discrete and specific somatosensory abnormalities, suggest that modification in peripheral and/or central nervous system processing may occur in BMS subjects and lend support to an organic cause for BMS in contrast to etiologies based primarily on psychogenic factors (see below).

COMMONLY SUGGESTED ETIOLOGIES

Psychological Dysfunction

Personality and mood changes (especially anxiety and depression) are clearly recognized as factors in BMS, often to the extent that BMS is viewed as a psychogenic condition.* Although current evidence clearly indicates a strong psychological component within BMS, there still exists no evidence of a close causal relationship between psychogenic factors and burning mouth as might be suggested through long-term prospective studies. Moreover, in considering a psychological explanation for the onset of BMS, it should also be noted that psychological dysfunction within a chronic pain population is common[77–80] and can occur as a result and not as a cause of chronic pain. In fact, it has been found that after successful treatment of chronic pain, patients often show a significant change to a more normal lifestyle,[78,81,82] whereas chronic pain patients who have been unsuccessfully treated may become suicidal.[82]

The conviction that a causal relationship does exist between psychogenic factors and the onset of BMS has been a major deterrent to the development of experimental models, other than psychogenic, to explain the pathogenesis of BMS and to further our understanding of the objective taste and somatosensory changes that accompany oral burning. Evidence for a psychogenic basis for BMS has been based both on psychological examination and on drug therapy studies in BMS subjects as outlined below.

Psychological Tests/Clinical Interviews

Numerous studies (both with and without control subjects) have used psychological questionnaires and/or psychiatric interviews to demonstrate psychological dysfunc-

---

*References 2, 10, 11, 12, 18–20, 22, 32, 48, 64–76.

tion (eg, depression, anxiety, obsessionalism, somatization, hostility) in subjects with oral complaints, including mouth burning.[†] Many of these reports (although not all: see, eg, Grushka et al[3] and Jontell et al[32]) claim to demonstrate a casual connection between psychogenic factors and BMS, but none, to date, has demonstrated more than a coincidental relationship. Even where no differences have been demonstrated between BMS and control subjects (eg, in one study by Eli et al[10] in which BMS patients and matched controls showed no significant differences in life events preceding the onset of oral burning), the investigators still interpreted this data to suggest that precipitation of BMS may be linked to the tendency among at-risk individuals for somatization, anxiety, and psychoticism following normal life events!

Another line of reasoning that has been used to label BMS as a psychogenic disorder originates from studies that have used tricyclic antidepressant (TCA) and benzodiazepine therapy in BMS. Harris and Davies,[68] for example, initially reported that subjects with BMS may suffer from cancerphobia or reactive depression. Studies[86] demonstrating that TCAs or benzodiazepines[2] were effective in alleviating the pain of BMS were used as further evidence that BMS was a psychogenic disorder. This conclusion, however, appears unwarranted because some of these studies,[86,87] as well as later ones,[88] also found that TCAs were as effective in treating "nonpsychiatric" or "nondepressed" as "psychiatric" or "depressed" subjects. Further, the studies that have used the benzodiazepines in unblinded trials[2,11] did not take into account other properties of benzodiazepines, including those related to their muscle relaxant effect.[89] This may be significant, since myofascial pain in the muscles of mastication has been found in some patients to be associated with mouth burning (eg, see Tourne et al[90]).

Menopausal Factors

It has long been suggested that hormonal changes at menopause may be a factor in the onset of oral discomfort,[91,92] based on the fact that most of the BMS subjects who present for treatment are postmenopausal. However, many studies of the effectiveness of ERT in postmenopausal BMS subjects[13,14,19,21] have not demonstrated a reduction of oral symptoms after the initiation of therapy. In contrast, Wardrop et al[23] recently reported reduced oral symptoms, including mouth burning, as well as decreased psychological factors in a majority of postmenopausal women who were treated with ERT; success was attributed to the alleviation by ERT of psychological distress, which then secondarily alleviated the oral discomfort. More recently, Forabosco et al[93] also demonstrated the effectiveness of hormone replacement therapy in alleviating symptoms of oral discomfort, including burning in 50% of affected subjects. These investigators additionally demonstrated a close correlation between those women who experienced symptom reduction and the presence of estrogen receptors in the oral mucosa. They concluded that this correlation may explain the diversity of responses observed among postmenopausal women to estrogen, and hence the discordant results among various studies.

[†] References 3, 6, 10, 16, 18, 20, 21, 48, 67, 72, 74, 83–85.

Nutritional Factors

Athough BMS has long been linked to nutritional deficiencies, especially of iron, B12, and folic acid,[34,94] many studies have failed to indicate a higher than expected prevalence of nutritional deficiencies in burning mouth.[4,23,29,52,95] Nevertheless it should be noted that there have been several recent reports of significantly greater deficiencies of vitamins B_1, B_2, B_6,[7,96] and zinc[97] in BMS subjects as compared to control subjects. However, replacement therapy of these vitamins[4,20] in some studies has produced resolution of symptoms in only 30% of deficient individuals[4] or else was no more effective than placebo.[20] Further, in a study by some of these same investigators,[98] no sustained (ie, greater than 2 months) improvement resulted from B_1, B_2, and B_6 replacement, even in those individuals diagnosed as deficient. Interestingly, in a more recent uncontrolled study by these same investigators,[7] B_1 and B_6 therapy alone, prescribed empirically in 312 individuals with burning mouth was found to be effective in reducing the intensity of burning pain by up to 45%. Unfortunately, no placebo-controlled data were reported, nor was the actual number of individuals who responded positively to this treatment. These findings have also not been reported by other investigators.

Diabetes Mellitus

Although diabetes mellitus has been implicated in BMS in a number of older studies,[99–101] the conclusions of some of these studies are suspect because of the very high prevalence of burning mouth in groups of supposedly normal dental patients and the high prevalence of abnormal glucose tolerance values in many individuals after treatment with oral hypoglycemic agents.[100] Recent findings suggest a prevalence of abnormal glucose tolerance tests in only 2% to 10% of diabetics[4,7,21,29] may also indicate that diabetes may not be an important cause of mouth burning. In contrast, however, diabetes may predispose to oral candidiasis, which may be responsible for oral burning in some patients (eg, see Tourne et al[90]).

Denture Allergy, Mechanical Irritation, Parafunctional Habits, and Galvanic Currents

High residual monomer levels in acrylic denture bases have been suggested as a cause of oral burning[48,102] either through allergy or chemical irritation. However, most recent studies[32,52,103–105] have not supported this hypothesis. A recent study has suggested instead that the mechanical irritation may be a more likely cause of burning associated with dentures. This supports other studies that have reported that errors in denture design as well as parafunctional habits (eg, tongue thrusting, grinding, clenching[4,5,14,21,30–32]) may be more important factors in BMS.

Galvanic currents were suggested almost 60 years ago as a cause of oral complaints, including mouth burning. However, recent controlled studies of patients with oral symptoms as well as asymtomatic controls[50,52–54,58] have shown that currents of similar intensity occur in both groups. In fact, a 1987 report by the American

Dental Association Council on Dental Materials, Instruments, and Equipment concluded that there is "no objective clinical method available for confirming decisively that galvanic currents in or between metallic dental restorations can cause subjective symptoms such as burning mouth, oral pain and stinging, battery, metal or salty tastes."[106]

Miscellaneous Causes

There have been multiple other suggestions put forward to account for BMS. Some of these include peanut sensitivity,[107] inflammation[108] or irritation[109] of the lymphoid tissue in the foliate papillae of the tongue, and an ischemia due to decreased blood supply to the tongue as a result of temporal or giant cell arteritis.[110] Other etiologies proposed include periodontal disease,[30,38] allergies,[50,98,111] oral candidiasis,[20,29,34] fusospirochetal infection,[112] myeloblastic syndrome,[113] sensory neuropathy,[114–116] reflux esophagitis,[90] geographic tongue,[2,29] acoustic nerve neuroma,[90] and myofascial referral sites from trigger points in the suprahyoid musculature (see Tourne et al[90] for review).

Management

TCA

Despite the variety of factors that have been suggested to account for BMS, treatment based on any one of these factors is often ineffective.[2,11,33] Tricyclic antidepressant medications, however, were initially used as therapeutic agents in BMS because of the belief that BMS was a psychogenic problem.[86] They have been found in a small number of clinical trials to be effective in alleviating idiopathic oral burning. Of note, these medications have been found to be as effective in low doses as they are at the higher doses usually used to treat depression, and to be effective in both "nondepressed" and "depressed" subjects with no associated mood change in the nondepressed subjects.[88]

The beneficial effects of the TCAs, including amitriptyline, desipramine, nortriptyline, imipramine, and clomipramine (see Max et al[117] for review), for relieving chronic pain have been corroborated in other recent studies that indicate that low doses of TCAs may act as analgesics, separate from their action as antidepressants.[117–119] Their mechanism of action is attributed to increased availability of synaptic neurotransmitters through inhibition of neronal reuptake.[117] On the basis of these types of studies, BMS is often managed with low doses of TCAs with good clinical results in some patients.[4,11,66,86–88,119] Reports of the greater efficacy of the benzodiazepines, eg, chlordiazepoxide and diazepam, as compared to the TCA amitriptyline[2] have not been confirmed in controlled double-blind, crossover studies.

Benzodiazepines

Other recent medications that have been reported anecdotally and in one uncontrolled retrospective study[120] to have some efficacy in ameliorating oral burning include clonazepam, a benzodiazepine sometimes used to treat trigeminal neuralgia and epilepsy.[121] Benzodiazepines, in general are GABA receptor agonists and promote brainstem serotonergic descending pain inhibition. They have been noted to suppress the spontaneous central neuronal hyperactivity that frequently occurs after differentiation. Clonazepam is thought to have a greater effect on the serotonin system in the brain and to have a longer half life than the other benzodiazepines.[122]

Miscellaneous/Others

Another recently reported medication that appears to have some effect in allevia-tion of burning mouth is topical capsaicin application.[123] Experimental studies have indicated that capsaicin can cause desensitization of the C-nociceptors of the polymodal type by first enhancing release of and then inhibiting reuptake of sub-stance P; prolonged treatment is thought to result in depletion of substance P by causing C-fiber degeneration.[124] In addition, capsaicin is thought to deplete many other neuropeoptides, presumably again as a consequence of degeneration.[124]

Many other medications and treatments have also been recommended for the symptomatic relief of the burning pain (for review, see Tourne and Fricton,[90] and Ship et al[37]), and others continue to be proposed, such as the systemic anesthetic mexiletine. These anesthetics are use-dependent sodium channel blockers[125] and have demonstrated efficacy in the treatment of neuropathic pain.[126] However, there is currently no information available with respect to its efficacy in BMS. The pro-posed mechanism of analgesic action for this class of drug includes suppression of spontaneous activity in damaged transmission in the spinal cord and stabilization of hyperexcitable nerve membranes centrally.

CONCLUSION

Although more is now known about BMS as a result of the increase in controlled studies, there still remains a great deal that is as yet unexplored. To further our knowledge of BMS and to better understand and manage patients, efforts must be directed at developing a working model of BMS that takes into account all currently available data. In this way, the model itself, or else parts of it, can be subject to objective scrutiny. Current studies suggest that the focus of future research on BMS should shift from its psychogenic aspects to elucidation of other organic and hope-fully treatable aspects of this disorder.

REFERENCES

1. Grushka M: Clinical features of burning mouth syndrome. Oral Surg 63:30–36, 1987.
2. Gorsky DMD, Silverman S Jr, Chinn H: Clinical characteristics and management outcome in the burning mouth syndrome. Oral Surg 72:192–195, 1991.
3. Grushka M, Sessle BJ, Miller R: Pain and personality profiles in burning mouth syndrome. Pain 28:155–167, 1987.
4. Lamey PJ, Lamb AB: Prospective study of aetiological factors in burning mouth syndrome. Br Med J 296:1243–1246, 1988.

5. Main DMG, Basker RM: Patients complaining of a burning mouth. Br Dent J 154:206–211, 1983.
6. van der Ploeg HM, van der Wal N, Eijkman MAJ, et al: Psychological aspects of patients with burning mouth syndrome. Oral Surg 63:664–668, 1987.
7. Lamey PJ, Lamb AB: Lip component of burning mouth syndrome. Oral Surg 78:590–593, 1994.
8. Grushka M, Katz RL, Sessle BJ: Spontaneous remission in burning mouth syndrome. J Dent Res 274, 1986.
9. Yontchev E, Hedegard B, Carlsson GE: Reported symptoms, diseases and medications of patients with orofacial discomfort complaints. Int J Oral Maxillofoc Surg 15:35–43, 1986.
10. Eli I, Kleinhauz M, Baht T, Littner M: Antecedents of burning mouth syndrome (glossodynia)—recent life events vs psychopathologic aspects. J Dent Res 73:567–572, 1994.
11. Gorsky DMD, Silverman S Jr, Chinn H: Burning mouth syndrome: A review of 98 cases. J Oral Med 42:7–9, 1987.
12. Hertz DG, Steiner JE, Zuckerman H, et al: Psychological and physical symptom-formation in menopause. Psychother Psychosom 19:47–52, 1971.
13. Pisanti S, Rafaely B, Polishuk WZ: The effect of steroid hormones on buccal mucosa of menopausal women. Oral Surg 40:346–353, 1975.
14. Ziskin DE, Moulton R: Glossodynia: A study of idiopathic orolingual pain. J Am Dent Assoc 33:1423–1432, 1946.
15. Grushka M, Sessle BJ: Burning mouth syndrome. Dent Clin North Am 35:171–184, 1991.
16. Zilli C, Brooke RI, Lau CL, et al: Screening for psychiatric illness in patients with oral dysaesthesia by means of the General Health Questionnaire—twenty-eight item version (GHQ-28) and the Irritability, Depression and Anxiety Scale (IDA). Oral Surg 67:384–389, 1989.
17. Grushka M, Sessls BJ: Taste impairment in burning mouth syndrome. Gerodontology 4:256–258, 1988.
18. Maresky LS, van der Bilj P, Gird I: Burning mouth syndrome. Evaluation of multiple variables among 85 patients. Oral Surg 75:303–307, 1993.
19. Ferguson MM, Carter J, Boyle P, et al: Oral complaints related to climacteric symptoms in oophorectomized women. J R Soc Med 74:492–498, 1981.
20. Browning S, Hislop S, Scully C: The association between burning mouth syndrome and psychosocial disorders. Oral Surg 64:171–174, 1987.
21. Basker RM, Sturdee DW, Davenport JC: Patients with burning mouths. A clinical investigation of causative factors, including the climacteric and diabetes. Br Dent J 145:9–16, 1978.
21. Lamey PJ, Lewis MAO: Oral medicine in practice: Burning mouth syndrome. Br Dent J 167:197–200, 1989.
22. Butlin HT, Spencer WG (eds): Diseases of the Tongue. London: Cassell and Co, 1900.
23. Wardrop RW, Hailes J, Burger H, Reade PC: Oral discomfort at menopause. Oral Surg 67:535–540, 1989.
24. Massler M: Oral manifestations during the female climacteric (the postmenopausal syndrome). Oral Surg 4:1234–1243, 1951.
25. Wolff A, Ship JA, Tylenda CA, Fox PG, Baum BJ: Oral mucosal appearance unchanged in healthy, different-aged persons. Oral Surg Oral Med Oral Pathol 71:569–572, 1991.
26. Locker D, Grushka M: The prevalence of oral and facial pain and discomfort: Preliminary results of a mail survey. Comm Dent Oral Epidemiol 15:169–172, 1987.
27. Klausner JJ: Epidemiology of chronic facial pain: Diagnostic usefulness in patient care. J Am Dent Assoc 125:1604–1611, 1994.
28. Gilpin SF: Glossodynia. JAMA 106:1722–1724, 1936.
29. Zegarelli DJ: Burning mouth: An analysis of 57 patients. Oral Surg 581:34–38, 1984.
30. Agerberg G: Signs and symptoms of mandibular dysfunction in patients with suspected oral galvanism. Acta Odontol Scand 45:41–48, 1987.
31. Haraldson T: Oral galvanism and mandibular dysfunction. Swed Dent J 9:129–133, 1985.
32. Jontell M, Haraldson T, Persson L, et al: An oral and psychosocial examination of patients with presumed oral galvanism. Swed Dent J 9:175–185, 1985.
33. Yontchev E, Hedegard B, Carlsson G: Outcome of treatment of patients with orofacial discomfort complaints. Int J Oral Maxillofac Surg 15:13–19, 1986.
34. Brooke RI, Seganski DP: Aetiology and investigation of the sore mouth. Can Dent Assoc J 10:504–506, 1977.
35. Grushka M, Shupak R, Sessle BJ: A rheumatological examination of 27 patients with burning mouth syndrome (BMS). J Dent Res 26:533, 1986.
36. Grushka M: Immunological aspects of "burning mouths." JAMA 249:1, 151, 1983.
37. Ship J, Grushka M, Lipton JA, Mott AE, Sessle BJ, Dionne RA: Burning mouth syndrome: An update. J Am Dent Assoc 126:843–853, 1995.
38. Johansson B, Stenman E, Bergman M: Clinical study of patients referred for investigation regarding so-called oral galvanism. Scand J Dent Res 92:469–475, 1984.

40. Pickett HG, Appleby RG, Osborn MO: Changes in the denture supporting tissues associated with the aging process. J Prosth Dent 27:257–262, 1972.
41. Baum BJ: Evaluation of stimulated parotid saliva flow rate in different age groups. J Dent Res 60:1292–1296, 1981.
42. Baum BJ: Salivary gland fluid secretion during aging. JAGS 37:453, 1989.
43. Ship JA, Patton LL, Tylenda CA: An assessment of salivary function in healthy premenopausal and postmenopausal females.
44. Tylenda CA, Ship JA, Fox PC, et al: Evaluation of submandibular salivary flow rate in different age groups. J Dent Res 67:1225, 1988.
45. Wu AS, Ship J: A characterization of major salivary gland flow rates in the presence of medications and systemic diseases. Oral Surg 76:301–306, 1993.
46. Persson RE, Izutsu KT, Truelove EL, Perrson R: Differences in salivary flow rates in elderly subjects using xerostomatic medications. Oral Surg Oral Med Oral Pathol 72:42–46, 1991.
47. Hugoson A: Results obtained from patients referred for the investigation of complaints related to oral galvanism. Swed Dent J 10:15–25, 1986.
48. Ali A, Reynolds AJ, Walker DM: The burning mouth sensation related to the wearing of acrylic dentures: An investigation. Br Dent J 161:444–447, 1986.
49. Glick D, Ben-Aryeh H, Gutman D, et al: Relation between idiopathic glossodynia and salivary flow rate and content. Int J Oral Surg 5:161–165, 1976.
50. Johansson B, Stenman E, Bergman M: Clinical registration of charge transfer between dental metallic materials in patients with disorders and/or discomforts allegedly caused by corrosion. Scand J Dent Res 94:357–363, 1986.
51. Syrjanen S, Piironen P, Yli-Urpo A: Salivary content of patients with subjective symptoms resembling galvanic pain. Oral Surg 58:387–393, 1984.
52. Axell T, Nilner K, Nilsson B: Clinical evaluation of patients referred with symptoms related to oral galvanism. Swed Dent J 7:169–178, 1983.
53. Bergman M, Ginstrup O, Nilner K: Potential and polarization measurement in vivo of oral galvanism. Scand J Dent Res 86:135–145, 1978.
54. Hampf G, Ekholm A, Salo T, et al: Pain in oral galvanism. Pain 29:301–311, 1987.
55. Yontchev E, Emilson G: Salivary and microbial conditions in patients with orofacial discomfort complaints. Acta Odontol Scand 44:215–219, 1986.
56. Navazesh M, Mulligan RA, Kipnis V, Denny PA, Denny PC: Comparison of whole saliva flow rates and mucin concentrations in healthy Caucasian young and aged adults. J Dent Res 71:1275–1378, 1992.
57. Närhi TO: Prevalence of subjective feelings of dry mouth in the elderly. J Dent Res 73:20–25, 1994.
58. Nilner K, Nilsson B: Intraoral currents of taste thresholds. Swed Dent J 6:105–113, 1982.
59. Gregg JM, Walter JR, Driscoll R: Neurosensory studies of trigeminal dysesthesia following peripheral nerve injury. In: Bonica J, Liebeskind JC, Albe-Fessard D (eds): Advances in Pain Research and Therapy, Vol. 3. New York: Raven Press, 1979.
60. Noordenbos W: Sensory findings in painful traumatic nerve lesions. In: Bonica J, Liebeskind JC, Albe-Fessard D (eds): Advances in Pain Research and Therapy, Vol. 3. New York: Raven Press, 1979.
61. Tasker RR, Tsuda T, Hawrylyshyn P: Clinical neurophysiological investigation of deafferentation pain. In: Bonica JJ, Lindblom U, Iggo A (eds): Advances in Pain Research and Therapy, vol. 4. New York: Raven Press, 1983.
62. Grushka M, Sessle BJ, Howley TP: Psychophysical assessment of tactile, pain and thermal sensory functions in burning mouth syndrome. Pain 28:169–184, 1987.
63. Svensson P, Bjerring P, Arendt-Nielsen L, Kaaber S: Sensory and pain thresholds to orofacial argon laser stimulation in patients with chronic burning mouth syndrome. Clin J Pain 9:207–215, 1993.
64. Engman MF: Burning tongue. Arch Derm Syphil 6:137–138, 1920.
65. Gerschman JA, Wright JL, Hall WD, et al: Comparisons of psychological and social factors in patients with chronic oro-facial pain and dental phobic disorders. Aust Dent J 32:331–335, 1987.
66. Goss A, McNamara J, Rounsefell B: Dental patients in a general pain clinic. Oral Surg 65:663–667, 1988.
67. Hampf G, Vikkula J, Yipaavalniemi P, et al: Psychiatric disorders in orofacial dysaesthesia. Int J Oral Maxillofac Surg 16:402–407, 1987.
68. Harris M, Davies G: Psychiatric disorders. In: Jones JH, Mason DK (eds): Oral Manifestations of Systemic Disease. London: WB Saunders, 1980.
69. Kutscher AH, Schoenberg B, Carr AC: Death, grief, and the dental practitioner: Thanatology as related to dentistry. J Am Dent Assoc 81:1373–1377, 1970.
70. Kutscher AH, Silvers HF, Stein G, et al: Therapy of idiopathic orolingual paresthesias. NY State J Med 52:1401–1405, 1952.

71. Lain ES: The mouth caused by artificial dentures. Arch Dermatol Syphil 25:21–32, 1932.
72. Lowenthal U, Pisanti S: The syndrome of oral complaints: Etiology and therapy. Oral Surg 46:2–6, 1978.
73. Massler M, Henry J: Oral manifestations during the female climacteric. Alpha Omegan 44:105–116, 1950.
74. Montgomery DW, Culver GD: Painful tongue. Arch Dermatol Syphil 26:474–477, 1932.
75. Sullivan P: The diagnosis and treatment of psychogenic glossodynia ENT. 68:1795–1798, 1989.
76. Hammarén M, Hugoson A: Clinical psychiatric assessmont of patients with burning mouth syndrome resisting oral treatment. Swed Dent J 13:77–88, 1989.
77. Sternbach R: Chronic pain as a disease entity. Triangle 20(1–2):27–32, 1981.
78. Sternbach RA, Timmermans G: Personality changes associated with the reduction of pain. Pain 1:177–181, 1975.
79. Turk DC, Rudy TE, Stein RL: Chronic pain and depression. Pain Management 2:17–25, 1987.
80. Watson D: Neurotic tendencies among chronic pain patients: An MMPI analysis. Pain 14:365–385, 1982.
81. Malow RM, Olsen RE: Changes in pain perception after treatment for chronic pain. Pain 11:65–72, 1981.
82. Shulman R, Turnbull IM, Diewald P: Psychiatric aspects of thalamic stimulation for neuropathic pain. Pain 13:127–135, 1982.
83. Lamb AB, Lamey PJ, Reeves PE: Burning mouth syndrome Psychological aspects. Br Dent J 165:256–260, 1988.
84. Lamey PJ, Lamb AB: The usefulness of the HAD scale in assessing anxiety and depression in patients with burning mouth syndrome. Oral Surg 67:390–392, 1989.
85. Rojo L, Silvestre FJ, Bagan JV, de Vicente T: Psychiatric morbidity in burning mouth syrdrome. Psychiatric interview versus depression and anxiety scales. Oral Surg 75:308–311, 1993.
86. Feinmann C, Harris M, Cawley R: Psychogenic facial pain: Presentation and treatment. Br Med J 288:436–438, 1984.
87. Dorey JL, Blasberg B, MacEntee MI, Conklin RJ: Oral mucosal disorders in denture wearers. J Prosthet Dent 53:210–213, 1985.
88. Sharav Y, Singer E, Schmidt E, Dionne RA, Dubner R: The analgesic effect of amitriptyline on chronic facial pain. Pain 31:199–209, 1987.
89. Carmen ME, Gillis MC, Bisson R: Compendium of Pharmaceuticals and Specialities, 28th ed. Toronto: Southam Murray, 1993.
90. Tourne LPM, Fricton JR: Burning mouth syndrome. Critical review and proposed clinical management. Oral Surg 74:158–167, 1992.
91. Nathanson IT, Weisberger DB: The treatment of leukoplakia buccalis and related lesions with estrogenic hormone. N Engl J Med 221:556–560, 1939.
92. Richman MJ, Abarbanel AR: Effects of estradiol, testosterone, diethylstilbesterol and several of their derivatives upon the human oral mucous membrane. J Am Dent Assoc 30:913–923, 1943.
93. Furabosco A, Criscuolo M, Coukos G, et al: Efficacy of hormonal replacement therapy in postmenopausal women with oral discomfort. Oral Surg 73:570–574, 1992.
94. Afonsky D: Deficiency glossitis. Oral Surg 4:482–500, 1951.
95. Bartoshuk LM, Rifkin B, Marks LE, et al: Taste and aging. J Gerontol 41:51, 1986.
96. Lamey PJ, Allam BF: Vitamin status of patients with burning mouth syndrome and the response to replacement therapy. Br Dent J 160:81–83, 1986.
97. Maragou P, Ivanyi L: Serum zinc levels in patients with burning mouth syndrome. Oral Surg. 71:447, 1991.
98. Lamey PJ, Lamb AB, Hughes A, Milligan KA, Forsyth A: Type 3 burning mouth syndrome: Psychological and allergic aspects. J Oral Pathol Med 23:216–219, 1994.
99. Cheraskin E, Brunson C, Sheridan RC, et al: The normal glucose tolerance pattern: The development of blood glucose normality by an analysis of oral symptoms. J Periodontol 31:123–137, 1960.
100. Chinn H, Brody H, Silverman S Jr, et al: Glucose tolerance in patients with oral symptoms. J Oral Ther Pharmacol 2:261–269, 1966.
101. Sheridan RC, Cheraskin E, Flynn FH, et al: Epidemiology of diabetes mellitis: II. A study of 100 dental patients. J Periodontol 30:298–323, 1959.
102. McCabe JE, Basker RM: Tissue sensitivity to acrylic resin. A method of measuring the residual monomer content and its clinical application. Br Dent J 140:347–350, 1976.
103. Kaaber S, Cramers M, Jepsen FL: The role of cadmium as a skin sensitizing agent in denture and non-denture wearers. Contact Dermatitis 8:308–313, 1982.
104. Yontchev E, Meding B, Hedegard B: Contact allergy to dental materials in patients with orofacial complaints. J Oral Rehab 13:183–190, 1986.
105. van Loon LAJ, Bos JD, Davidson CL: Clinical evaluation of fifty-six patients referred with symptoms tentatively related to allergic contact stomatitis. Oral Surg 74:572–575, 1992.

106. Council on Dental Materials, Instruments, and Equipment: American Dental Association status report on the occurrence of galvanic corrosion in the mouth and its potential effects. J Am Dent Assoc 115:783–787, 1987.
107. Whitley BD, Holmes AR, Shepherd MG, Ferguson MM: Peanut Sensitivity as a cause of burning mouth. Oral Surg 72:671–674, 1991.
108. Sluder G: A case of glossodynia with lingual tonsillitis as its etiology: Control through the nasal ganglion. JAMA 81:115, 1923.
109. Makienko MA: Treatment of glossalgia. Dent Abs 3:378, 1958.
110. Norman JE: Facial pain and vascular disease. Br J Oral Surg 8:138–144, 1970.
111. James J, Ferguson MM, Forsyth A: Mercury allergy as a cause of burning mouth. Br Dent J 159:392, 1985.
112. Katz J, Benoliel R, Leviner E: Burning mouth sensation associated with fusospirochettal infection in edentulous patients. Oral Surg 62:152–154, 1986.
113. Gilson J, Lamey PJ, Watson WH: The myeloblastic syndrome presenting with oral symptoms. Br Dent J 163:234–235, 1987.
114. Bachrach WH: Painful tongue. Dent Abstr 8:48, 1963.
115. Itkin AB: The entrapment syndrome. J NJ State Dent Soc 40:29–35, 1968.
116. Kutscher AH, Chilton NW: Dolometric evaluation of idiopathic glossodynia. NY State Dent J 18:31–32, 1950.
117. Max MB, Kishore-Kumar R, Schafer SC, et al: Efficacy of desipramine in painful diabetic neuropathy: A placebo-controlled trial. Pain 45:3–9, 1991.
118. Baraldi M, Poggiol R, Sandi M, et al: Antidepressants and opiates interactions: Pharmacological and biochemical evidences. Pharm Res Commun 15:843–857, 1983.
119. Singer E: Pain control in dentistry: Management of chronic orofacial pain. Compend Cont Dent Ed 114:116–118, 1987.
120. Hampf G, Aalberg V, Sunden B: Experiences from a facial pain unit. J Craniomand Dis Fac Oral Pain 4:267–272, 1990.
121. Zakrzewska JM, Patsalos PN: Drugs used in the management of trigeminal neuralgia. Oral Surg 74:439–450, 1992.
122. Sellers EM, Khanna JM: Anxiolytics, hypnotics, and sedatives. In: Kalant H, Roschlau WHE (eds): Principles of Medical Pharmacology, 5th edition. BC Decker, 1989, pp. 255–264.
123. Epstein JB, Marcoe JH: Topical application of capsaicin for treatment of oral neuropathic pain and trigeminal neuralgia. Oral Surg 77:135–140, 1994.
124. Lynn B: Capsaicin: Actions on nociceptive C-fibers and therapeutic potential. Pain 41:61–69, 1990.
125. Catterall WA: Common modes of drug action on Na^+ channels: Local anaesthetics, antiarrhythmics and anticonvulsants. Trends Pharmacol Sci 8:57–65, 1987.
126. Strache H, Myer UE, Schumacher HE, Federlin K: Mexiletine in the treatment of diabetic neuropathy. Diabetes Care 15:1550–1555, 1992.

13

Aging and the Chemical Senses

CLAIRE MURPHY, Ph.D.
JILL RAZANI, M.A.
TERENCE M. DAVIDSON, M.D.

Elderly patients presenting to a physician with complaints about taste or smell present a unique challenge to the clinician. It is likely that in addition to potentially acute etiologies for their chemosensory dysfunction, aging alone has taken its toll. It is important to be aware of age-associated losses and to ascertain the degree to which patients suffer from such loss. It is equally important not to dismiss unexplained chemosensory loss as simply attributable to age.

Precipitous loss of chemosensory function is cause for concern. Studies have demonstrated impairment in smell function in older individuals at the threshold and suprathreshold intensity levels, as well as on tasks such as odor identification and odor memory that require cognitive functioning. Although deficits associated with aging have been shown for taste, this change is of lesser magnitude when compared to olfaction. In addition, it appears that the age-related changes are not uniform for the four taste qualities of sweet, sour, salty, and bitter. In this chapter we review information on age-associated chemosensory dysfunction in the hope of aiding the clinician in the interpretation of clinical assessment of taste and smell in the older patient and, thus, in the differential diagnosis of age-associated loss and loss that is attributable to other causes (head trauma, viral illness, nasal/sinus disease, etc). The chapter concludes with information on effective clinical assessment of taste and smell in the older patient.

The senses of taste and smell work in concert to produce chemosensory experiences and perceptions. What one experiences as flavor in food, for example, is the product of olfaction as well as taste and is enhanced and/or suppressed by other sensations such as heat, cooling, etc. When the chemical senses are functioning normally, the chemosensory systems respond to stimulation with a certain degree of sensitivity and accuracy. Head trauma, sinus disease, viral illness, toxic exposures, medication, oral disease, and aging contribute to chemosensory dysfunction: decreased sensitivity or intensity, and distortion of tastes and smells. Thus, chemosensory changes can result from a combination of conditions that may include aging, disease, medication, and environment.

TASTE

It is common for elderly patients to complain about dullness of their taste sensitivity. Studies examining age-related changes in taste function have produced equivocal results.[1-4] These results may be due to differences in methodology, discrepancies between threshold and suprathreshold functioning, and subjects' health status. Despite such discrepancies, important information can be gleaned from such studies.

Taste Threshold

The taste threshold is the lowest concentration at which a person can just perceive a stimulus. Since sensitivity shows moment-to-moment fluctuation, the threshold is often stated as the lowest concentration a subject will detect on a given percentage of trials. Virtually all studies examining taste threshold have shown a modest decline in sensitivity associated with aging.[1,2,4-6] Although decreased taste sensitivity for sucrose,[7-13] sodium chloride,[7,9,11,14] citric acid,[10] quinine,[9,14-18] 6-*n*-propylthiouracil,[16] and phenylthiocarbamide[19,20] have been reported for the elderly, the uniformity of these changes for the four taste qualities (sweet, salty, sour, and bitter) still remains a subject of discussion.

In an earlier study by Murphy,[5] seven males' and seven females' detection thresholds for sucrose, NaCl, citric acid, tartaric acid, HCl, caffeine, quinine sulfate, and magnesium sulfate were measured, and results revealed that sensitivity varied as a function of age, tastant, and an age-by-tastant interaction. Quality-specific age effects have also been demonstrated by other investigators. In an investigation of age-associated taste sensitivity for the four taste qualities using the stimuli sucrose, NaCl, quinine sulfate, and citric acid, Weiffenbach, Baum, and Burghauser[14] demonstrated decreased sensitivity for NaCl and quinine sulfate in the elderly, but no statistically significant age-associated differences in sucrose and citric acid. Another interesting finding of this study was the gender effect for citric acid, with males showing less sensitivity than females. In a more recent study, Weiffenbach, Cowart, and Baum[21] again showed similar results for sucrose and NaCl. A scatter plot of their data revealed that NaCl thresholds increased as a function of age, while the slope for sucrose thresholds remained flat. Spitzer[22] also tested representatives of the taste qualities sweet, sour, bitter, and salty and found age-associated losses in the latter three taste qualities, but not in sucrose. The consensus of the literature seems to be that the taste threshold for some taste qualities (salty and bitter) is affected more in old age than others (sweet). There is also some debate about the magnitude of the decrease in threshold sensitivity as well as its role in the perception of real-world stimuli.

Recent data confirm the hypothesis that age-associated changes in the taste system may also differ for different compounds within a taste quality, particularly bitter.[23] Schiffman et al[17] compared taste thresholds for 12 different bitter compounds in young and elderly subjects. For 6 out of the 12 compounds, threshold sensitivity was poorer for the elderly, and the least lipophilic compounds showed the greatest age-related loss. These results argue for compound-specific decline in bitter sensitivity with advancing age. Similarly Cowart, Yokomukai, and Beauchamp[15] examined threshold and suprathreshold sensitivity for two bitter compounds, urea and quinine sulfate, and reported loss of sensitivity with age for quinine sulfate but not for urea. This study lends additional support for the existence multiple bitter taste transduction mechanisms in humans and the idea that these mechanisms may be differentially affected by age. Schiffman and colleagues'[24] observation that age-related losses in detection threshold for 10 sodium salts were greatest for salts with anions having the greatest molar conductivity further suggests that differential losses at taste threshold are dependent on differences at the transduction level.

Suprathreshold Intensity

Further support for alterations in the taste system due to aging is provided by suprathreshold intensity studies. With the aid of recent psychophysical techniques such as magnitude estimation and magnitude matching it is now possible to investigate the ability to distinguish among and track increases in varying stimulus concentrations (see chapter 10). With magnitude matching, subjects are required to estimate the magnitude of a series of stimuli from one modality as well as from another modality, all on the same scale of magnitude.[25] Magnitude estimation is different in that subjects are required to assign numbers reflecting stimulus intensity to a series stimuli that vary in concentration, but must do so without an external anchor.[26] Thus, both methods can provide insight into whether an older person perceives appropriate increases in intensity with increases in stimulus concentration. Intensity estimates plotted as a function of concentration result in a psychophysical function, the slope of which reflects this relationship. In addition, magnitude matching provides information about the relative intensity of a given taste for a young person as compared to an older person. In this case, the intercept of the function will be meaningful.

A number of studies of age-associated changes in taste intensity have been conducted using the method of magnitude estimation. Enns et al[27] reported no changes in the slopes of the psychophysical functions for sucrose from young adulthood to old age. Bartoshuk et al[28] confirmed this. They also interpreted the slopes for sour and salty to be generally stable in the elderly, but noted some flattening near the threshold level, particularly for the bitter stimulus. Other investigators, reporting age-associated effects on the slopes of psychophysical functions for taste substances, lend support to this picture of greater impairment in the ability of the elderly to track increases in stimulus concentration for some substances.[29–32] The study by Schiffman and Clark[29] was particularly intriguing because they investigated amino acids, most of which are generally bitter, and reported that slopes were flatter for elderly persons. Interestingly, those amino acids that were complexed with HCl showed steeper slopes.

Analyzing a different aspect of magnitude estimation data, the intraclass correlation, Weiffenbach, Cowart, and Baum[21] also found impaired performance in the elderly, with significant decrements for bitter, salty, and sour, but not sweet. Proportion of total variance accounted for by the variation in stimulus strength in subjects' judgment ratings is represented in the intraclass correlation coefficients. Thus, the repeatability of judgments within a testing session is expressed in this coefficient. The result: Judgments of the elderly were less reliable.

Murphy and Gilmore,[23] using magnitude matching, assessed perceived intensity of the tastants sucrose, caffeine, sodium chloride, and citric acid, relative to ratings of weight. Ratings of weight are reasonably stable over the life-span. Intensity ratings for the following two-component mixtures were also assessed: sucrose/caffeine, sucrose/citric acid, and sucrose/sodium chloride. The results showed that the elderly subjects found bitter and sour substances to be significantly less intense than the young, but no significant differences in intensity ratings for sweet and salty unmixed components were found (see Fig. 13–1). In addition, based on the effect sizes, age-associated loss in bitterness was greater (20% of the variance accounted for by age) than that in sourness (10% of the variance accounted for by age). One of the more interesting findings of this study was that the results indicated a quality-specific intensity loss in the mixture components when rated in the mixture. The elderly subjects judged bitter to be less intense in the sweet-bitter mixture than the young, whereas saltiness of sweet-salty and sourness of sweet-sour mixtures remained relatively stable across the ages (Fig. 13–2). Thus, the elderly subjects' intensity judgments for bitter were significantly lower than those of the young

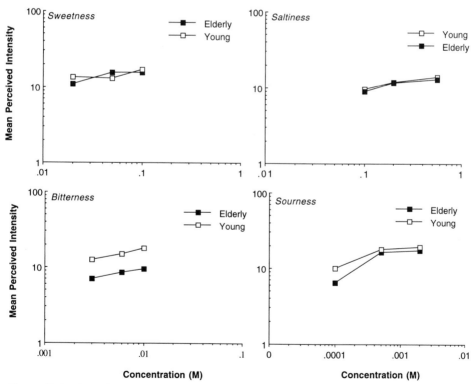

Figure 13–1. Suprathreshold intensity estimates generated by young and elderly subjects for low, medium, and high concentrations of sweet, salty, and bitter stimuli. Data replotted from Murphy and Gilmore.[23]

subjects for both the unmixed and the mixed components. Clearly, bitterness perception seems to be most affected in the elderly.

Cowart[33] also examined suprathreshold scaling for 137 subjects ranging in age from 19 to 87, with at least 19 subjects represented in each decade of life, for sucrose, NaCl, citric acid, and quinine sulfate. She found a significant main effect of age for all taste qualities. However, the effects of age were greatest for the perception of intensity for the bitter and sour taste stimuli.

Discrimination ability of young and elderly individuals for sucrose and caffeine has also been assessed by measuring Weber ratios.[23] The ability to discriminate

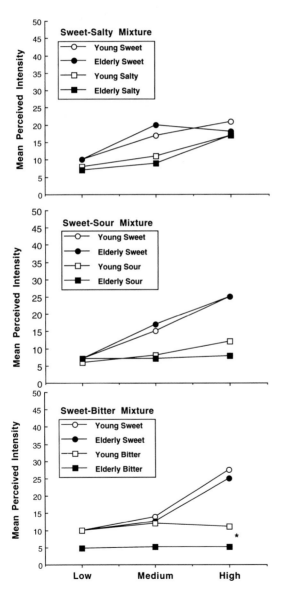

Figure 13–2. Suprathreshold intensity estimates generated by young and elderly subjects for two-component mixtures of low, medium, and high concentrations of sweet, salty, sour, and bitter stimuli. The elderly subjects judged bitter to be less intense than the young in the sweet-bitter mixture, as indicated by the asterisk. Data replotted from Murphy and Gilmore.[23]

between a given concentration of a stimulus and a standard concentration is measured in terms of the just noticeable difference (JND). The ratio of the JND to a particular standard concentration results in a Weber ratio (WR). In this particular study the investigators were interested in two specific questions: (1) whether elderly individuals produce greater WRs than the young subjects, ie, require greater increases in concentration to detect differences in stimuli, and (2) whether the size of the WR would be dependent upon the stimulus tested. The results revealed no differences between the WRs for the young and elderly for sucrose at any of the standard concentration levels, suggesting that the same magnitude of change in stimulus concentration is needed for both groups to perceive a noticeable difference. However, differences were found between the WRs of the young and elderly for caffeine. The greatest difference, a WR of 2.5, was found for the elderly at the highest concentration level, suggesting that a 250% change in concentration level was required for the elderly to be able to perceive a noticeable difference. In contrast, a WR of 0.25 was observed for the young, suggesting that only a 25% increase in concentration was necessary to perceive a difference.

In a subsequent study, Nordin et al[34] employed the method of constant stimuli for measuring WRs in young and elderly to investigate age-related loss in JNDs for citric acid and NaCl. Each subject also performed an absolute threshold task to ensure that all subjects were able to taste the range of the stimulus concentrations. The results of this study once again supported the hypothesis that age-related changes in intensity perception are quality-specific. Results indicated differences in WRs between young and elderly for citric acid, but not for NaCl. In comparing effect size for bitter and sour qualities, it was found that the effect size was larger for caffeine (0.17) than for citric acid (0.16).

The above discussion is relevant to the clinician testing taste function in older persons. A true picture of function will require testing taste response to stimuli of more than one quality and at more than one level of the system. Losses in bitter detection, intensity, and discrimination can more accurately be ascribed to age-associated causes than losses in other qualities.

OLFACTION

Threshold

As stated previously, the perception of food flavor involves input from olfaction as well as from taste. The olfactory input courses retronasally to stimulate the olfactory system and thus adds additional chemosensory stimulation to the taste of food. Hence, loss of olfactory function will significantly impact flavor perception in the elderly.

Impairment of human olfactory function as a result of aging has been reported at the level of absolute threshold by a number of investigators, using a variety of measures and stimuli.[1–4,35–42]

Past research in the field has employed a variety of populations to assess age effects on olfactory function. Clinical work over the past decade has highlighted the incidence of olfactory impairment as a result of nasal inflammatory disease, eg, allergic rhinitis; bacterial rhinosinusitis[43]; postviral hyposmia; and head trauma.[44] A number of dementias (eg, Alzheimer's disease [AD] and Parkinson's disease, as discussed elsewhere in this chapter), which tend to be most common in later life, are also associated with olfactory impairment. Furthermore, even elderly persons who have not met the criteria for the diagnosis of dementia and who would remain unidentified in the general population but who, on the basis of a standard neuropsychological scale, would qualify for the diagnosis of questionable dementia or "at risk" for dementia show impairment in olfactory function.[45]

We have found in ongoing work[47] that the effects of aging itself on olfactory threshold have been somewhat overestimated in earlier studies, including some of our own, because elderly subjects were included without regard to their nasal health and without regard to their dementia status. When the criteria for participation included normal nasal health and absence of dementia, and butanol thresholds were assessed using a two-alternative, forced-choice procedure, the initial baseline data showed that the subjects tended to be less sensitive as a function of age, as we and others have reported in previous studies.[1,3,33,37,48–51] However, the difference in olfactory function between older and younger subjects prescreened for nasal health and absence of dementia is substantially smaller than when subjects are not screened. Figure 13–3 compares thresholds for an unscreened elderly population (although active, living independently and able to travel) and a prescreened elderly population (see Fig. 13–2).

Thus, the elderly patient presenting to the physician with olfactory loss should be viewed in the context created by current work on the nonimpaired elderly and evaluation of patients' odor threshold performances should be evaluated against correct age-appropriate standards.

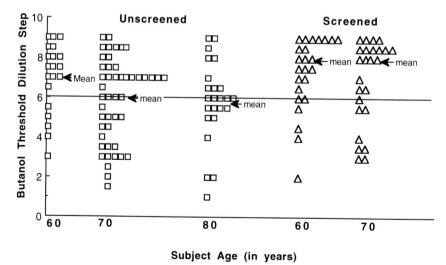

Figure 13–3. Thresholds seen in 60-, 70-, and 80-year-old subjects prescreened for dementia and nasal disease compared with thresholds for persons where screening for nasal health was not an option. Higher dilution steps, ie higher numbers, correspond to better thresholds. Group means are indicated by arrows.

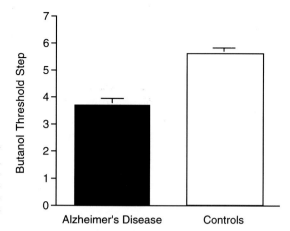

Figure 13–4. Mean (± SE) olfactory thresholds for butanol in patients with probable Alzheimer's disease and in age-matched controls. Higher dilution steps indicate better function. The average score for the Alzheimer's group indicates moderate to severe hyposmia.

Problems with olfactory function at threshold levels are especially pronounced in patients with AD, who account for approximately 55% of residents of nursing homes. Patients with AD show neuropathological changes characteristic of the disease (eg, neuritic plaques and neurofibrillary tangles) in olfactory areas of the brain: anterior olfactory nucleus, entorhinal cortex, and piriform cortex.[52] It has also been reported that there are histopathological changes in the olfactory epithelium of patients who died with the disease.[53] Reports of odor identification deficits were the first to suggest olfactory functional involvement in the disease.[54,55] Because AD patients have significant deficits in semantic memory, naming of odors might be particularly difficult, regardless of odor sensitivity. However, studies of olfactory threshold have also revealed significant deficits in these patients, even when using a two-alternative forced-choice method (to control for criterion problems) and employing a control task (assessing thresholds for sucrose, a test with similar task demands) to rule out problems with performance of a threshold task.[56]

Figure 13–4, which compares odor thresholds for AD patients and controls participating in our ongoing study of olfactory dysfunction in Alzheimer's disease, illustrates the nature of the effect. Taste thresholds for sucrose, a task included to rule out the possibility that any differences we might see in olfactory function could be ascribed to cognitive deficits rather than sensory deficits, revealed no differences between normal elderly and Alzheimer's patients.

We have evaluated the rhinologic status of normal elderly persons and patients with AD. This study was motivated by: (1) psychophysical data[54–56] showing functional deficits in the olfactory system, which correlated with the degree of dementia[56]; (2) the significant amount of neuropathological evidence to suggest that patients with AD have olfactory system abnormalities[53,57,58]; and (3) the fact that rhinologic status of this group had not been characterized in spite of the potential for peripheral nasal disease to account for the olfactory dysfunction observed in these patients.

Olfactory function was assessed by threshold test. Rhinologic examination included rhinomanometry before and after topical application of the decongestant neosynephrine; nasal cytograms graded for eosinophils, neutrophils, basophils, epithelial cells, goblet cells, and bacteria, and endoscopic examination for signs of infection (eg, purulent rhinorrhea), inflammatory disease (eg, polyps), and anatomical derangement (eg, septal deviation). The olfactory cleft was examined and the olfactory epithelium was evaluated when accessible. The status of the olfactory mucosa was staged as normal, abnormal, or not seen. Alzheimer's disease patients had significantly poorer olfactory thresholds than normal, age-matched controls. Alzheimer's patients and controls showed no difference in prevalence of nasal disease, irrespective of whether the clinical examination including endoscopy, the patient history, or the secondary diagnosis based on cytological findings was used to assign diagnoses. Thus, we reported in Feldman et al[59] that the findings support a neurologically mediated phenomenon as the cause for significant impairment in olfactory function in AD.

An additional observation in this study has potential clinical significance. State of the olfactory epithelium was graded as normal or abnormal based on appearance on endoscopic examination. Normal epithelium was raised compared with the flatter surrounding respiratory epithelial surfaces. The epithelium was considered abnormal in the case of absent mucosa, creating a scarred or denuded cleft, or a thin layer of respiratory mucosa replacing olfactory mucosa. Interestingly, patients who had abnormal olfactory epithelium on gross endoscopic examination had significantly poorer olfactory sensitivity, regardless of dementia status. Thus, physical changes in the peripheral olfactory system (olfactory epithelium) were found to accompany olfactory deficits in the elderly. Whether this represents peripheral insult (eg, as a result of viral infection) or degeneration as a result of loss of central connections (as seen in head trauma) is an interesting empirical question.

Persons with Down's syndrome who live to be 40 years of age show Alzheimer's-like changes in the brain: neuritic plaques, neurofibrillary tangles, and loss of choline acetyltransferase. Chromosome 21 is implicated in both abnormalities.[60] Murphy and Jinich[61] investigated whether older Down's syndrome patients (average age = 30 years) show olfactory dysfunction, using a series of olfactory tasks: odor threshold, odor identification with the University of Pennsylvania Smell Identification Test (UPSIT),[62] odor identification with a nonlexical odor identification test,[45,63] and odor recognition memory.[64]

Olfactory thresholds in the Down's syndrome subjects were significantly higher than those of the controls, in spite of normal performance in a taste threshold task, a task similar to the olfactory threshold in terms of subject demands. Down's syndrome patients were also significantly impaired on the odor identification test, scoring approximately half the number typical of normals. Odor recognition memory was also significantly impaired. Thus, older Down's syndrome patients exhibited significant impairment, whether assessed by odor threshold measurement, odor identification, or odor recognition memory. Results suggest the potential for olfactory deficits to provide a sensitive indicator of the deterioration and progression of brain pathology in older persons with Down's syndrome. They also alert the clinician to the need to view olfactory performance from the perspective of dementia when assessing chemosensory status in the older Down's patient.

Other dementias common in the elderly have also been associated with olfactory deficits. For example, symptoms of Huntington's disease (HD) tend to develop later in life. Moberg et al[65] reported loss of odor memory in HD. We recently investigated odor detection sensitivity, intensity discrimination, quality discrimination, odor identification, and recognition memory in persons in their 50s with HD and found significant loss of olfactory function.[66] Impairment was greatest in odor identification, and losses were significant for odor threshold (see Figs. 13–4 and 13–5). Our findings agreed with the observations of Doty[67] and further indicate that poor threshold sensitivity in HD cannot be explained by impaired task comprehension, nor impaired odor identification by lexical difficulties. Losses in discrimination were associated with detection deficits. Results suggest the importance of awareness of the impact of HD on olfactory function in clinical patients, since they exhibit significant olfactory dysfunction, particularly in odor detection sensitivity and odor identification, the two measures most commonly used in clinical assessment.

Several studies have indicated that both threshold testing and identification show deficits in patients with Parkinson's disease.[68–70] Both medicated and unmedicated patients are affected.[67] Doty et al[70] have argued that the deficit is unrelated to disease severity or disease duration. Thus, older patients with Parkinson's disease should be evaluated for chemosensory status with this deficit in mind.

The importance of olfactory loss for safety issues for the elderly population can not be overstated. Chalke and Dewhurst,[71] who examined threshold for town gas, revealed that the recognition threshold for town gas was two to five times greater for individuals over the age of 65 than in people younger than 65. Thus, the average older person has a threshold for the warning odor in gas that renders him or her significantly less able to use that warning odor for protection from the dangers of leaking gas. Similarly, a later study by Stevens et al[72] reported the average elderly threshold for ethyl mercaptan, the odorant in liquified petroleum gas, to be 10-fold higher than that of the young.

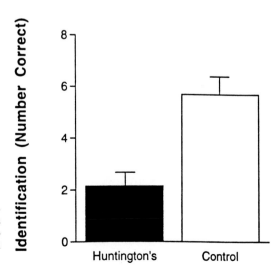

Figure 13–5. Mean (± SE) odor identification scores for patients with Huntington's chorea indicate severe dificits relative to age-matched controls. Data from Nordin et al.[66]

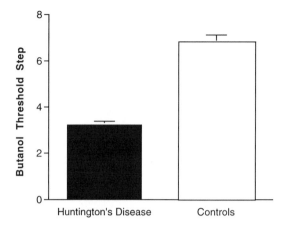

Figure 13–6. Mean (± SE) olfactory thresholds for butanol in patients with Huntington's chorea and age-matched controls. Higher-dilution steps indicate better threshold sensitivity. The Huntington's patients' average performance indicates severe hyposmia. Data replotted from Nordin et al.[66]

A major issue for those elderly persons who have a diminished sense of smell is awareness of the smell impairment. Nordin et al[73] set out to examine the degree to which an elderly individual is able to recognize and assess the severity of his or her smell dysfunction. Eighty sinusitis patients, 80 AD patients, and 80 nondemented elderly controls performed a detection threshold task and reported their own perception of their smell function on a questionnaire. The results revealed that 77% of the elderly, 74% of the AD patients, and 8% of the sinusitis patients with smell loss reported normal smell sensitivity. These findings emphasize the need for education hyposmic and anosmic individuals about their smell loss and the potential dangers that accompany it. For example, patients should be adequately counseled regarding their inability to detect the following at normal levels: smoke, leaking gas, and spoiled food.

These data also speak to the issue of the prevalence of smell loss in the US population and efforts to estimate prevalence using self-report instruments or numbers of physician visits with a primary complaint of anosmia or hyposmia. Such approaches will tend to underestimate smell impairment in the elderly. Questionnaire data may be no substitute for actual sensitivity testing of samples of the population to gain accurate information on the prevalence of smell loss in the US population. An accurate estimate of the prevalence of smell impairment in the US population will require actual sensitivity testing of samples of the population as well as accurate information about sinonasal disease, age, dementia status, environmental exposure, gender, and genetic factors.

Suprathreshold Intensity

Both magnitude estimation and magnitude matching have demonstrated significant effects of aging on suprathreshold scaling of odor intensity.[37,74] Murphy[37] examined perceived intensity judgments, using the method of magnitude estimation,[26] in young and elderly participants for the odorant menthol across a number of concentrations. The results showed that the median slope of the psychophysical functions generated by the older subjects was half that of the younger group, indicating that the ability to track increases in stimulus concentration was significantly impaired in the elderly. Of the concentrations in the series, the elderly perceived fewer, on average, than the young. Employing the method of magnitude matching, Stevens and Cain[74] also examined intensity judgments for odors in elderly and young subjects and found that elderly subjects rated each of six odorants as less intense than the younger subjects. Further, the slope of the function for the elderly was smaller than that of the young for two (benzaldehyde and ethyl alcohol) of the six odorants.

These studies of odor intensity characterize the decrease in odor intensity ratings of elderly individuals relative to the young. However, they do not inform about the degree to which individuals require an increase in stimulus concentration to perceive a difference (ie, JND) in odor sensation. Investigations of JNDs and WRs, the ratios of JNDs to standard concentration,[75] in the young and elderly address these questions.

In a recent study, Razani and Murphy[76] examined WRs for odor in young and elderly adults to determine whether WRs for odor stimuli are greater for elderly than for young subjects and whether WRs vary differentially with stimulus concentration in the two age groups. The elderly produced greater WRs than the young, ie, the elderly required a greater difference in concentration between the two stimuli to notice a change in intensity. Interestingly, there was a greater difference in WRs between the young and elderly at the lowest standard concentration, perhaps because this concentration was closest to the elderly's absolute threshold.

Results of these psychophysical studies make it clear that for older people, perception of odors is less intense, and it is more difficult to discriminate between intensities of the same odor. This will be true for odors such as smoke, gas, and perfume, as well as those associated with food preparation and the appreciation of food flavors.

Odor Identification

In a seminal study, Schiffman and Pasternak[77] compared the ability of young and elderly adults to identify puréed foods. The elderly exhibited a striking inability to identify the flavors. Murphy[78,79] later confirmed this and went on to demonstrate that the older person's difficulty with food identification was driven more by impairment in olfaction than impairment in taste. When deprived of olfaction, younger people performed at the same level as the older people. A cognitive effect also became salient: with practice all subjects improved their ability to identify the foods; however, elderly persons benefited less from feedback regarding the accuracy of their labels than younger subjects did. Supplying the correct label on each trial (presumably enhancing semantic encoding) improved retrieval of that label on subsequent trials to a greater degree for younger subjects than for older subjects.

Other studies examining the elderly's ability to identify nonfood odors also indicate impairment. Schemper et al[80] assessed the ability of young and elderly women to discriminate among 20 common odors. Subjects who performed well (80% accuracy) were later asked to identify 40 additional odors. The elderly performed poorer than the young. Subsequent testing with the 40-item odor battery indicated that the older subjects benefited neither from odor labels presented to them by the experimenters nor from labels they produced. These results suggested less active encoding or spontaneous use of verbal mediators in elderly persons.

A study performed by Doty et al[81] demonstrated poor odor identification skills in a large sample of the elderly. In this study, a sample of 1955 subjects from 5 to 99 years of age were given the UPSIT, a self-administered, 40-item, microencapsulated, four-alternative, forced-choice odor identification test. More than half of the elderly subjects in this study had seriously compromised olfactory function, with individuals over the age of 80 performing poorest. Although virtually all studies examining age-associated changes in odor identification ability have shown a decline with age, the magnitude of the effect is still under discussion. A recent study by Ship and Weiffenbach,[82] using the UPSIT, showed smaller odor identification deficits in the elderly. One explanation for this difference may be the relative health status of the elderly in the two studies.

Murphy and Cain[83] examined odor identification for 80 odors in sample of 40 persons ranging in age from 19 to 66 years and found a linear decline over the 45-year age span. Half of the participants were blind and half were sighted. The overall results of the study indicated that although the blind were no more sensitive than the sighted, ie, their thresholds were no better, they had more highly developed odor identification skills. Thus, it appeared that the blind who had a need for identifying objects through a modality other than sight had made particularly good use of their olfactory abilities. The differential effects of aging on the two groups were telling: the negative correlation with age ($r = -0.81$, $P < 0.001$ for the sighted) was less in the blind ($r = -0.73$, $P < 0.001$). Although we can only speculate that the overlearning of odor labels in the blind helped to spare them from some of the decline of age, these results are certainly consistent with other studies of cognitive function in the elderly that show a protective effect of education and professional use.

Wysocki and Gilbert[84] used the National Geographic Smell survey, a scratch-and-sniff 6-item smell test, to evaluate odor identification. Evaluating the results from surveys returned by 1.2 million people, presumably readers of National Geographic magazine and/or their family members, they found ability to identify the odors to be significantly associated with age. Older females showed better performance than males, but both groups showed losses. The average percent correct fot those young adult subjects able to detect each odor ranged from approximately 20% to 90% over all odors. Losses ranged from approximately 10% to over 30%.

A series of studies[55,85–90] have shown that AD patients have difficulty identifying odors, whether from free recall of names or in a recognition format with odor names from which to choose the correct identification (UPSIT). Even very early-stage AD patients show the loss.[45] Figure 13–7 illustrates the deficit in odor identification in patients with the designation "at risk" for dementia, based on performance on neuropsychological testing and neurological examination. The contributions of olfactory sensory loss and loss of cognitive function to breakdown in performance on

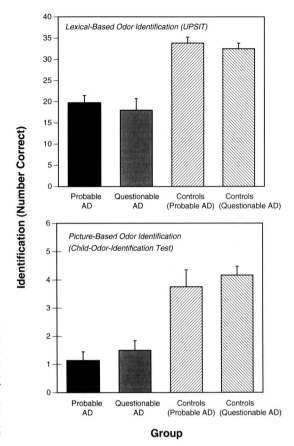

Figure 13–7. Mean number correct (± SE) for odor identification in probable Alzheimer's disease (AD) patients, questionable AD patients, and their respective control groups. The upper graph illustrates performance on the University of Pennsylvanis Smell Identification Test, and the lower graph illustrates performance on the Child Odor Identification Test. Data replotted from Morgan et al.[45]

this test have not been quantified; however, the implications for clinical testing with odor identification tests are clear: results from such a patient must be interpreted in the context of performance of other AD patients.

Odor Memory

Recognition memory is the method that has typically been used for assessing memory for odors. One of the most common odor recognition memory paradigms is one in which target odors are presented to the subject and, after a retention period, the subject is again presented with several or all of the target stimuli as well as several distractor odors. Memory is then assessed by calculating percent correct judgments or some derivative measure based on accuracy. Young subjects typically retain and accurately recognize a large number of the original odors presented under this paradigm. In one of the earlier studies, young subjects correctly recognized about 70% of the odors even after a 2-month retention period.[91] When tested in such an odor memory test, the elderly showed large decrements in performance.[42] The fall-off in performance occurred as early as 15 minutes after presentation and performance continued to deteriorate over a 2-week time period. Six months after presentation the young continued to perform well above chance on recognition memory for odors presented in this paradigm. The elderly performed as if they had never been exposed to the odors. Differences in odor threshold sensitivity accounted for part of the decline in odor memory. Correlations of memory performance with identification ability suggested that older people were less able to use naming of the odors as an aid to remember their presentation. Hence, it appeared that difficulty in encoding the odors played a significant role in memory difficulty.

In the elderly, memory for visual and auditory stimuli is poorer for recall, which requires one to conjure up the information without a cue, than for recognition, in which some cue is provided as an aid.[92–95] In a recent study of odor recall, recognition, and identification, the elderly showed significant deficits in the ability to recall odors and verbally presented words, including immediate recall as well as short- and long-delay recall, both free and cued by use of semantic categories.[96]

In contrast to the young, the elderly showed limited use of semantic-clustering strategies that would have supported and facilitated semantic encoding. There were no differences in serial clustering. Whereas some investigators have reported minor effects of semantic encoding on odor memory,[97–99] others have reported rather significant effects. It has, for example, been demonstrated that the availability of appropriate verbal labels greatly enhances odor-recognition memory,[100–102] and that suppression of verbal encoding suppresses odor recognition.[42,103] The above discussion would suggest that performance of an elderly person in an odor task will be poorer for recall than for recognition where a cue is presented to aid retrieval from memory.

Event-Related Potentials

Electroencephalogram and event-related potential studies have been used productively in clinical settings to evaluate auditory, visual, and somesthetic function. Development of the olfactory event-related potential for application in clinical testing has lagged behind, largely because of difficulties with adaptation during the presentation of large numbers trials at relatively short intervals, the small amplitude of odor-evoked responses, and stimulus control. However, Kobal and Hummel[104] have repeatedly applied the technique to good effect in both the trigeminal realm and in the olfactory realm. Older people, particularly those with mild or significant cognitive impairment, are difficult to assess with some psychophysical methods because of subject bias, criterion shifts, and memory problems. Because of its promise as an objective, noninvasive measure of sensory function, assessment of olfactory function with the olfactory event-related potential (OERP) in the elderly is attractive. Murphy et al[105] recorded OERPs in young and elderly persons. Results indicated that amplitude of the OERP correlated with psychophysical odor threshold and that older subjects showed decreased amplitude and increased latency of N1/P2 of the OERP. Thus, these measurements of brain activity in response to stimulation support psychophysical findings of olfactory functional impairment in the elderly. The development of the OERP makes available a powerful tool to better characterize and probe the processes that underlie impaired olfactory function in the elderly.

CLINICAL ASSESSMENT OF CHEMOSENSORY FUNCTION

Evaluation of the older patient requires sensitivity to the existence of age-associated loss and its potential to be superimposed on impairment from causes that can be treated with current medical and surgical interventions.

The history is particularly important in the assessment of the elderly. Certain diseases that are more common in the elderly are associated with significant losses in olfactory function, eg, AD, Huntington's chorea, Parkinson's disease, and older patients with Down's syndrome. Assessment of cognitive function in these patients will be important for proper interpretation of performance on odor identification testing, a task that relies heavily on cognitive function. Loss of semantic memory in AD patients will render this test very difficult for such a patient, regardless of his or her ability to detect the presence of odors. Even threshold measurement will be correlated with degree of dementia in AD, and thus assessment of nasal/sinus-related olfactory dysfunction will be particularly challenging.

Assessment of Taste Function

Although taste complaints are less common than olfactory complaints, patients present with complaints of ageusia (complete taste loss), hypogeusia (partial loss of taste function), and dysgeusia (a persistent, unpleasant, foul taste in the mouth). Assessment of taste function in the elderly patient is complicated by the need to assess all four taste qualities, since, as described in detail above, age-associated loss is of greater magnitude for one quality than for another. Our approach to assessment at UCSD includes estimation of intensity and pleasantness and identification of stimuli of all four qualities. This approach is capable of detection taste loss for stimuli above threshold (and, presumably, of greater consequence to the patient), as well as probing for the presence of dysgeusia. See Murphy et al[105] for an approach that employs spatial testing of different tongue regions and Cowart[33] for the use of threshold testing in clinical populations, as well as chapter 10.

Magnitude matching is used to assess suprathreshold intensity functions and estimate the magnitude of suprathreshold loss. Patients estimate the intensity of a series of weights (50, 100, 200, 400, and 500 g) intermixed with the series of four concentrations each of sucrose (0.05, 0.10, 0.20, 0.40 mol/L), NaCl (0.05, 0.10, 0.20, 0.40 mol/L), citric acid (0.0006, 0.0012, 0.0024, 0.0048 mol/L), and caffeine (0.0025, 0.005, 0.01, 0.02 mol/L). The resultant estimates of intensity for taste are examined for the ability to track increases in stimulus concentration, as well as for absolute intensity levels relative to weights. Weights, rather than auditory cues, represent the matching continuum in our assessments because of the large number of elderly patients who have hearing loss. Estimates of pleasantness are obtained at the same time and will be altered in the dysgeusic patient. Identification of the tastants depends on sensitivity as well as absence of dysgeusia; thus, misidentifications of the less-concentrated stimuli typically reflect loss, while misidentifications characterizing pleasant stimuli (eg, sucrose) as unpleasant suggest dysgeusia.

In patients complaining of dysgeusia, an estimate of the intensity of the persistent unpleasant taste is obtained prior to conducting the above assessment and again at its conclusion. The patient is given a topical anesthetic and then asked at 2-minute intervals to rate the dysgeusia over the next 10 minutes. In between these estimates, the most concentrated of each of the four taste stimuli presented in the above assessment (0.4 mol/L sucrose, 0.4 mol/L NaCl, etc) are given again so that the patient rates these tastes after the topical anesthetic. The data from the dysgeusia test are then examined to assess the effect of the anesthetic on the four taste stimuli and on the patient's report of dysgeusic taste. Dysgeusia of peripheral origin typically is reduced drastically by the anesthetic, as are the taste estimates.

Thus, information regarding taste intensity, identification of common taste substances, and reports of pleasantness and unpleasantness as well as estimates for dysgeusic tastes combine to produce a profile of chemosensory function in the patient with a taste complaint. Because of the potential for sinonasal disease to present as a taste complaint in patients with olfactory dysfunction—particularly dysgeusia secondary to sinusitis, or in combination with olfactory dysfunction—olfactory assessment is often necessary to complete the chemosensory assessment.

Electrogustometry has been widely used in Japan for examining oral function in a clinical setting. Murphy et al[106] explored the advantages and limitations of electrogustometry for assessment of taste dysfunction in young and elderly adults. They measured thresholds for electrical stimulation delivered by a commercially available device, using a two-alternative, force-choice, ascending method, and assessed free and cued descriptors for the quality of the electrical stimulus at threshold. Threshold sensitivity was significantly lower for the elderly. Test-retest reliability was high, and thresholds for left and right sides of the tongue were highly correlated. Thresholds for electrical stimulation were not correlated with thresholds for aqueous solutions of citric acid or NaCl, stimuli that evoke sour and salty sensations. When the thresholds for the electrical stimulus were compared to WRs for NaCl and citric acid in the elderly, the two measures showed no correlation. Salata et al[107] showed a similar lack of correspondence between electrical stimulation and perception of intensity of chemical solutions on the tongue. Subjects' reports of sensory quality suggested that the sensations evoked by electrical stimulation at threshold intensity level are not well represented by the descriptors *sweet*, *salty*, *sour*, and *bitter*. Only 3 of the 22 subjects freely described the sensation as a "basic" taste quality, and in the cued descriptor condition, only 4 of the 22 reported a basic taste quality; however, buzzing vibration was reported correctly by 18 of the 22 subjects. Thus, the study raises improtant questions regarding the peripheral mechanisms underlying electrical stimulation, whether electrical stimulation results in taste (or vibrotactile or trigeminal) sensations, and whether the puntate stimulation with electrical stimulation can be a reliable and valid measure for assessment of the type of taste loss complaint with which patients typically present to the clinic.

Certainly it suggests the merit of further investigation of electogustometry before adoption as a method acceptable for diagnosis of taste dysfunction in the clinical setting.

Assessment of Olfactory Function

An olfactory assessment typically includes both threshold measurement and assessment of suprathreshold function, the latter often in the form of an odor identification test. In our hands the threshold test of sensitivity to butanol is a two-alternative, forced-choice test with an interstimulus interval of 45 seconds.[56,108] A forced-choice threshold test is important for patients of any age, but particularly for the elderly whose criterion for reporting the presence of the odor may be altered either in an effort to be absolutely sure of the presence of dorant or simply to please the experimenter. The latter problem is magnified in the demented patient. In our clinical experience, the elderly patient is more susceptible to the influences of adaptation, and thus the controlled interstimulus interval in particularly important for an accurate assessment of ability. In interpreting the results of threshold testing it is important to be aware of the expected degreee of age-associated loss. Some elderly patients rationalize loss of chemosensory function with the thought that everything is affected by aging and that loss of smell and taste is expected to be extensive and pervasive in old age. Recent work indicates that the degree of olfactory loss in the healthy elderly person is less than reported in older studies of institutionalized elderly patients whose dementia status and health, particularly nasal sinus health, was unknown.[2] It is important that losses due to treatable causes do not go unnoticed or ignored in the elderly due to a mistaken notion that they are expected. Precipitous, significant loss of function is always cause for investigation, regardless of the age of the patient. Since elderly persons typically underestimate a gradual loss of function,[109] any indication of impairment is a particular cause for concern.

Odor identification ability is an integral part of olfactory assessment in all patients, but the method should be appropriate for the patient. Odor identification for a battery of common odors, using an intertrial interval of 45 seconds and a cue sheet to aid memory, is appropriate for most adults. Assessment of children and of others with reading of semantic difficulties is carried out at the UCSD Nasal Dysfunction Clinic using an odor identification test with no lexical demands. As odors are presented for identification, an array of pictures is presented and the patient is asked to point to the picture that corresponds to the odor presented. We have found this test to be reliable in children as young as 5 or 6 years of age.[63] Administration of the UPSIT is particularly appropriate for patients prevention to the clinic who are involved in litigation, since the test scoring includes a category for probable malingering.[110]

Odor identification is influenced both by the ability to sense odors and the ability to name them. The latter is particularly susceptible to dementing diseases and to the influence of normal aging. Thus, scores on odor identification tests for elderly patients would be expected to be lower than those for the young, apart from sensory deficits. Interpretation of results should therefore be tempered by a knowledge of the age of the patient.

Counseling

Counseling the patient with olfactory loss is important at any age, but particularly in the elderly. Patients need to be aware of the limitations imposed by their disability and cautioned regarding their safety. Because the patient will not detect the odor of leaking gas or smoke at normal levels, precautions need to be taken (eg, purchasing smoke alarms, gas detectors for gas appliances). They must also be counseled about spoiled and rotten foods. An aggressive leftover discard policy is advised. Asking a normosmic individual to taste foods is also recommended. Older people may need support regarding feelings of self-consciousness about their inability to appreciate fine food and wine when among friends or when preparing food. Some patients may be concerned about bodily odors, particularly halitosis and flatulence, that they cannot detect. Information and reassurance that their problems are real, understood, and shared by others is of real value to these patients.

Acknowledgments—Preparation of this chapter was supported by NIH Grant # AG04085 from the National Institute on Aging. We gratefully acknowledge the contributions of the members of the Life-Span Human Senses Laboratroy, San Diego State University and of the Alzheimer's Disease Research Center of the University of California, San Diego to the research described in this chapter.

REFERENCES

1. Murphy C: Taste and smell in the elderly. In: Meiselman HL, Rivlin RS (eds): Clinical Measurement of Taste and Smell. New York: Macmillan, 1986.
2. Murphy C: Nutrition and chemosensory perception in the elderly. Crit Rev Food Sci Nutr 33:3–15, 1993.
3. Schiffman SS: Age-related changes in taste and smell and their possible causes. In: Meiselman HL, Rivlin RS (eds): Clinical Measurement of Taste and Smell. New York: Macmillan, 1986.
4. Schiffman SS: Perception of taste and smell in elderly persons. Crit Rev Food Sci Nutr 33:17–26, 1993.
5. Murphy C: The Effects of age on taste sensitivity. In Han SS, Coons DH (eds): Special Senses in Aging. Ann Arbor, MI: University of Michigan Institute of Gerontology, 1993.
6. Schiffman SS: Changes in taste and smell with age: Psychophysical aspects. In: Ordy, Brizzee (eds): Sensory Systems and Communication in the Elderly. New York: Raven Press, 1979.
7. Balogh K, Lelkes K: The tongue in old age. Gerontol Clin 3(suppl):38–54, 1961.
8. Bouliere F, Cendron H, Rappatort A: Modification avec l'lage des seuils gustatifs de reconnaissance aux saveurs salee et surcre, chez l'homme. Gerontologia 2:104–112, 1977.
9. Cooper RM, Bilash I, Zubek JP: The effect of age on taste sensitivity. J Gerontol 14:56–58, 1959.
10. Hermel J, Schonwetter S, Samueloff S: Taste sensation and age in man. J Oral Med 25:39–42, 1970.
11. Hinchcliffe R: Clinical quantitative gustometry. Acta Otolaryngol 49:453–466, 1958.
12. Richter C, Campbell K: Sucrose taste thresholds of rats and humans. Am J Physiol 128:291–297, 1940.
13. Schiffman SS, Lindley MG, Clark TB, Makins C: Molecular mechanism of sweet taste: Relationship of hydrogen bonding to taste sensitivity in both young and elderly. Neurobiol Aging 2:173–185, 1981.
14. Weiffenbach JM, Baum BJ, Burghauser R: Taste thresholds: Quality specific variation with human aging. J Gerontol 37:700–706, 1982.
15. Cowart BJ, Yokomukai Y, Beauchamp GK: Bitter taste in aging: Compound-specific decline in sensitivity. Physiol Behav 56:1237–1241,1994.
16. Glanville EV, Kaplan AR, Fischer R: Age, sex, and taste sensitivity. J Gerontol 19:474–478, 1964.
17. Schiffman SS, Gatlin LA, Frey AE, Heiman SA, Stagner WC, Cooper DC: Taste perception of

bitter compounds in young and elderly persons: Relation to lipophilicity of bitter compounds. Neurobiol Aging 15:743–750, 1994.

18. Smith SE, Davies PP: Quinine taste thresholds: A family study and a twin study. Ann Hum Genet 37:227–232, 1973.
19. Harris H, Kalmus H: The measurement of taste sensitivity to phenylthiourea (PTC). Ann Hum Genet 15:24–31, 1949.
20. Kalmus H, Trotter WR: Direct assessment of the effect of age on PTC sensitivity. Ann Hum Genet 26:145–149, 1962.
21. Weiffenbach JM, Cowart BJ, Baum BJ: Taste intensity perception in aging. J Gerontol 4:460–468, 1986.
22. Spitzer ME: Taste acuity in institutionalized and noninstitutionalized elderly men. J Gerontol Psychol Sci 43:71–74, 1988.
23. Murphy C, Gilmore MM: Quality-specific effects of aging on the human taste system. Percept Psychophys 45:121–128, 1989.
24. Schiffman SS, Clark CM, Warwick ZS: Gustatory and olfactory dysfunction in dementia: Not specific to Alzheimer's disease. Neurobiol Aging 11:597–600, 1990.
25. Marks LE, Stevens JC: Measuring sensation in the aged. In: Poon LW (ed): Aging in the 1980's: Psychological Issues. Washington, DC: American Psychological Association, 1980.
26. Stevens SS: On the psychopysical law. Psychol Rev 64:153–181, 1957.
27. Enns MP, Van Itallie TB, Grinker JA: Contributions of age, sex and degree of fatness on preferences and magnitude estimation for sucrose in humans. Physiol Behavi 22:999–1003, 1979.
28. Bartoshuk LM, Rifkin B, Marks LE, Bars P: Taste and aging. J Gerontol 41:51–57, 1986.
29. Schiffma SS, Clark TB: Magnitude estimates of amino acids for young and elderly sujects. Neurobiol Aging 1:81–91, 1980.
30. Schiffman SS, Lindley MG, Clark TB, Makins C: Molecular mechanism of sweet taste: Relationship of hydrogen bonding to taste sensitivity in both young and elderly. Neurobiol Aging 2:173–185, 1981.
31. Hyde RJ, Feller RP: Age and sex effects on taste of sucrose, NaCl, citric acid and caffeine. Neurobiol Aging 2:315–318, 1981.
32. Cowart BJ: Direct scaling of the intensity of basic tastes: A life span study. Fifth annual meeting of the Association for Chemoreception Sciences, Sarasota, FL, 1983.
33. Cowart BJ: Relationships between taste and smell across the adult life span. In: Murphy C, Cain WS, Hegsted DM (eds): Nutrition and the Chemical Senses in Aging. New York: New York Academy of Sciences, 1989.
34. Nordin S, Razani LJ, Markison S, Murphy C: Quality-specific age-related effects on absolute and different taste thresholds for citric acid and NaCl. Percept Psychophys (accepted for publication).
35. Kimbrell GM, Furchtgott E: The effect of aging on olfactory threshold. J Gerontol 18:364–365, 1963.
36. Venstrom D, Amoore JE: Olfactory threshold in relation to age, sex or smoking. J Food Sci 33:264–265, 1968.
37. Murphy C: Age-related effects on the threshold, psychophysical function, and pleasantness of menthol. J Gerontol 38:217–222, 1983.
38. Schiffman SS, Hornak K, Reilly D: Increases taste thresholds of amino acids with age. Am J Clin Nutr 32:1622–1627, 1981.
39. Schiffman SS: Age-related changes in taste and smell and their possible causes. In: Meiselman HL, Rivlin RS (eds): Clinical Measurement of Taste and Smell. New York: Macmillan, 1986.
40. Van Toller C, Dodd GH, Billing A: Aging and the Sense of Smell. Springfield, IL: CC Thomas, 1987.
41. Stevens JC, Cain WS, Schiet FT, Oatley MW: Aging speeds olfactory adaptation and slows recovery. Ann NY Acad Sci 561:323–325, 1989.
42. Murphy C, Cain WS, Gilmore MM, Skinner RB: Sensory and semantic factors in recognition memory for odors and graphic stimuli: Elderly versus young persons. Am J Psychol 104:161–192, 1991.
43. Mott AE: Topical corticosteroid therapy for nasal polyposis. In: Getchell T, Doty RL, Bartoshuk LM, Snow JB (eds): Smell and Taste in Health and Disease. Raven Press, 1991.
44. Costanzo RM, Zasler ND: Epidemiology and pathophysiology of olfactory and gustatory dysfunction in head trauma. J Head Trauma Rehabil 7:15–24, 1991.
45. Morgan, CD, Nordin S, Murphy C: Odor identification as an early marker for Alzheimer's disease: Impact of lexical functioning and detection sensitivity. J Clin Exp Neuropsychol 17:1–11, 1995.
46. Nordin S, Murphy C: Impaired sensory and cognitive olfactory function in questionable Alzheimer's disease. Neurpsychology 10:113–119, 1996.
47. Murphy C, Nordin S, de Wijk R, Cain WS, Polich J: Development of olfactory evoked potentials for functional assessment in the elderly. Chem Senses 8:605, 1993.

48. Murphy C: Aging and chemosensory perception of and preference for nutritionally significant stimuli. Nutrition and the chemical senses in aging: Recent advances and current research needs. Ann NY Acad Sci 561:251–266, 1989.

49. Murphy C: Nutrition and chemosensory perception in the elderly. Crit Rev Food Sci Nutr 33:3–15, 1993.

50. Schiffman SS: Perception of taste and smell in elderly persons. Crit Rev Food Sci Nutr 33:17–26, 1993.

51. Stevens JC, Cain WS: Old-age deficits in the sense of smell as gauged by thresholds, magnitude matching and odor identification. Psychol Aging 2:36–42, 1987.

52. Price JL, Davis DB, Morris JC, White DL: The distribution of tangles, plaques and related immunohistochemical markers in healthy and aging and Alzheimer's disease. Neurobiol Aging 12:295–312, 1991.

53. Talamo BR, Rudel R, Kosik KS, et al: Pathological changes in olfactory neurons in patients with Alzheimer's disease. Nature 337:736–739, 1989.

54. Serby M: Olfactory deficits in Alzheimer's disease. J Neural Transmission 24:69–77, 1987.

55. Doty RL, Reyes P, Gregor T: Olfactory dysfunction in Alzheimer's disease. Brain Res Bull 18:597–600, 1987.

56. Murphy C, Gilmore MM, Seery CS, Salmon DP, Lasker BR: Olfactory thresholds are associated with degree of dementia in Alzheimer's disease. Neurobiol Aging 11:465–469, 1990.

57. Esiri M, Wilcock G: The olfactory bulbs in Alzheimer's disease. J Neurol Neurosurg Psychiatry 47:56, 1984.

58. Reyes PF, Golden GT, Fagel PL, Fariello RG, Katz L, Carner E: The prepiriform cortex in dementia of the Alzheimer's type. Arch Neurol 44:644–645, 1987.

59. Feldman JI, Murphy C, Davidson TM, Jalowayski AA, Galindo de Jaime G: The rhinologic evaluation of Alzheimer's disease. Laryngoscope 101:1198–1202, 1991.

60. Delabar JM, Goldgaber D, Lamour Y, et al: Amyloid gene duplication in Alzheimer's disease and karyotypically normal Down syndrome. Science 226:1441–1392, 1981.

61. Murphy C, Jinich S: Olfactory dysfunction in Down's syndrome. J Gerontol (in press).

62. Doty RL, Shaman P, Applebaum SL, Gilverson R, Sikorsky L, Rosenberg L: Smell identification ability: Changes with age. Science 226:1441–1443, 1984.

63. Anderson J, Maxwell L, Murphy C: Odorant identification testing in the young child. Chem Senses 17:5, 1992.

64. Murphy C, Cain WS, Gilmore MM, Skinner RB: Sensory and semantic factors in recognition memory for odors and graphic stimuli: Elderly versus young persons. Am J Psychol 104:161–192, 1991.

65. Moberg PJ, Pearlson GD, Speedie LJ, Lipsey JR, Strauss ME, Folstein SE: Olfactory recognition: Differential impairments in early and late Huntington's and Alzheimer's disease. J Clin Exp Neuropsychol 9:650–664, 1987.

66. Nordin S, Paulsen JS, Murphy C: Sensory- and memory mediated olfactory dysfuntion in Huntington's disease. J Int Neuropsychol Soc 1:281–290, 1995.

67. Doty RL: Olfactory dysfunction in neurodegenerative disorders. In: Getchell TV, Doty RL, Bartoshuk LM, Snow JB Jr (eds): Smell and Taste in Health and Disease. New York, Raven Press, 1991.

68. Ansari KA, Johnson A: Olfactory function in patients with Parkinson's disease. J Chronic Dis 28:493–497, 1975.

69. Ward CD, Hess WS, Calne DB: Olfactory impairment in Parkinson's disease. Neurology 33:943–946, 1983.

70. Doty RL, Deems DA, Stellar S: Olfactory dysfunction in Parkinson's: A general deficit unrelated to neurologic signs, disease stage, or disease duration. Neurology 38:1237–1244, 1987.

71. Chalke HD, Dewhurst JR: Accidental coal-gas poisoning. Br Med J 2:915–917, 1957.

72. Stevens JC, Cain WS, Weinstein DE: Aging impairs the ability to detect gas odor. Fire Tech 23:198–204, 1987.

73. Nordin S, Monsch A, Murphy C: Unawareness of smell loss in normal aging and Alzheimer's disease: Discrepancy between self-reported and diagnosed smell sensitivity.

74. Stevens JC, Cain WS: Old-age deficits in the sense of smell as gauged by thresholds, magnitude matching, and odor identification. Psychol Aging 2:36–42, 1987.

75. Engen T: Psychophysics: I. Discrimination and detection. In: Kling JW, Riggs LA (eds): Woodworth & Schlosberg's Experimental Psychology. New York: Holt, Rinehart and Winston, 1971.

76. Razani J, Nordin S, Murphy C: Differences in odor intensity discrimination in young and elderly. J Gerontol (submitted).

77. Schiffman SS, Pasternak M: Decreased discrimination of food odors in the elderly. J Gerontol 34:73–79, 1979.

78. Murphy C: Effects of aging on food perception. J Am Coll Nutr 1:128–129, 1982.
79. Murphy C: Cognitive and chemosensory influences on age-related changes in the ability to identify blended foods. J Gerontol 40:47–52, 1985.
80. Schemper T, Voss S, Cain WS: Odor identification in young and elderly persons: Sensory and cognitive limitations. J Gerontol 18:446–452, 1981.
81. Doty RL, Shaman P, Applebaum SL, Giberson R, Siksorski L, Rosenberg L: Smell identification ability: Changes with age. Science 226:1441–1443, 1984.
82. Ship, Weiffenbach: Age, gender, medical treatment, and medical effects on smell identification. J Gerontol 48:M26–M32, 1993.
83. Murphy C, Cain WS: Odor identification: The blind are better. Physiol Behav 37:177–180, 1986.
84. Wysocki CJ, Gilbert AN: National Geographic Smell Survey: Effects of age are heterogenous. Ann NY Acad Sci 510:12–28, 1989.
85. Serby MJ, Corwin J, Conrad P, Rotrosen RP: Olfactory dysfunction in Alzheimer's disease and Parkinson's disease. Am J Psychiatry 142:781–782, 1985.
86. Cowin J, Serby M, Conrad P, Conrad P, Rotrosen J: Olfactory recognition deficit in Alzheimer's and Parkinsonian dementias. IRCS Med Sci 13:260, 1985.
87. Knupfer L, Spiegel R: Differences in olfactory test performance between normal aged, Alzheimer and vascular type dementia individuals. Int J Geriat Psychiatry 1:3–14, 1986.
88. Warner MD, Peabody CA, Flattery JJ, Tinklenberg JR: Olfactory deficits and Alzheimer's disease. Biol Psychiatry 21:116–118, 1986.
89. Koss E: Olfactory dysfunction in Alzheimer's disease. Dev Neuropsychol 2:89–99, 1986.
90. Rezek DL: Olfactory deficits as a neurologic sign in dementia of the Alzheimer type. Arch Neurol 44:1030–1032, 1987.
91. Lawless H: Recognition of common odors, pictures and simple shapes. Percept Psycophys 24:493–495, 1978.
92. Botwinick, Storandt: Memory Related Functions and Age. Chicago, IL: Charles C. Thomas, 1974.
93. Erber J: Age differences in recognition memory. J Gerontol 29:177–181, 1974.
94. Harwood E, Naylor GFK: Recall and recognition in elderly and young subjects. Aust J Psychol 21:251–257, 1969.
95. Rainowitz CA, Craik FI, Ackerman BP: A processing resource account of age differences in recall. Can J Psychol 36:323–344, 1982.
96. Murphy C, Nordin S, Acosta L: Odor recall in young and elderly adults: Processes, strategies, constituents and comparison to word recall. Neuropsychology (accepted for publication).
97. Engen T, Ross BM: Long-term memory odors with verbal descriptions. J Exp Psychol 100:221–227, 1973.
98. Lawless HT, Cain WS: Recognition memory for odors. Chem Senses Flavor 1:331–337, 1975.
99. Lawless HT, Engen T: Associations to odors: Interference, memories, and verbal labeling. J Exp Psychol 3:52–59, 1977.
100. Lyman BJ, McDaniel, MA: Effects of encoding strategy on long-term memory for odours. Q J Exp Psychol 38A:753–765, 1986.
101. Rabin MD, Cain WS: Odor recognition, familiarity, identifiability, and encoding consistency. J Exp Psychol Learning Memory Cogni 10:316–325, 1984.
102. Walk HA, Johns EE: Interference and facilitation in short-term memory for odors. Percept Psychophys 36:508–514, 1984.
103. Perkins J, McLaughlin Cook N: Recognition and recall of odours: The effects of suppressing visual and verbal encoding processes. Br J Psychol 81:221–226, 1990.
104. Kobal G, Hummel T: Olfactory evoked potentials in humans. In: Getchell TV, Doty RL, Bartoshuk LM, Snow JB (eds): Smell and Taste in Health and Disease. New York: Raven Press, 1991.
105. Murphy C, Nordin S, de Wijk RA, Cain WS, Polich J: Olfactory-evoked potentials: Assessment of young and elderly, and comparison to psychophysical threshold. Chem Senses 19:47–56, 1994.
106. Murphy C, Quinonez C, Nordin S: Reliability and validity of electrogustometry and its application to young and elderly persons. Chem Senses 10:499–503, 1995.
107. Salata JA, Raj JM, Doty RL: Differential sensitivity of tongue areas and palate to electrical stimulation: A suprathreshlod cross-modal matching study. Chem Senses 16:483–489, 1991.
108. Cain WS, Gent JS, Catalanotto FA, Goodspeed RB: Clinical evaluation of olfaction. Am J Otolaryngol 4:252–256, 1983.
109. Nordin S, Murphy C, Davidson TM, Quinonez C, Jalowayski AA, Ellison DW: Prevalence and assessment of qualitative olfactory dysfunction in different age groups. Laryngoscope (accepted for publication).
110. Doty RL, Shaman P, Dann M: Development of the University of Pennsylvania Smell Identification Test: A standardized microencapsulated test of olfactory function. Physiol Behavi 32:489–502, 1984.

Index